Applications of
Time Series Analysis
to Evaluation

Garlie A. Forehand, *Editor*

NEW DIRECTIONS FOR PROGRAM EVALUATION
A Publication of the Evaluation Research Society
SCARVIA B. ANDERSON, *Editor-in-Chief*

Number 16, December 1982

Paperback sourcebooks in
The Jossey-Bass Higher Education and
Social and Behavioral Sciences Series

Jossey-Bass Inc., Publishers
San Francisco • Washington • London

Applications of Time Series Analysis to Evaluation
Number 16, December 1982
      Garlie A. Forehand, *Editor*

New Directions for Program Evaluation Series
A Publication of the Evaluation Research Society
Scarvia B. Anderson, *Editor-in-Chief*

New Directions for Program Evaluation (publication number
USPS 449-050) is published quarterly by Jossey-Bass Inc.,
Publishers, and is sponsored by the Evaluation Research Society.
Second-class postage rates paid at San Francisco, California,
and at additional mailing offices.

Correspondence:
Subscriptions, single-issue orders, change of address notices,
undelivered copies, and other correspondence should be sent to
New Directions Subscriptions, Jossey-Bass Inc., Publishers,
433 California Street, San Francisco, California 94104.

Editorial correspondence should be sent to the Editor-in-Chief,
Scarvia B. Anderson, Educational Testing Service, 250 Piedmont
Avenue, Suite 2020, Atlanta, Georgia 30308.

Library of Congress Catalogue Card Number LC 81-48580
International Standard Serial Number ISSN 0164-7989
International Standard Book Number ISBN 87589-918-8

Cover art by Willi Baum
Manufactured in the United States of America

Ordering Information

The paperback sourcebooks listed below are published quarterly and can be ordered either by subscription or as single copies.

Subscriptions cost $35.00 per year for institutions, agencies, and libraries. Individuals can subscribe at the special rate of $21.00 per year *if payment is by personal check.* (Note that the full rate of $35.00 applies if payment is by institutional check, even if the subscription is designated for an individual.) Standing orders are accepted.

Single copies are available at $7.95 when payment accompanies order, and *all single-copy orders under $25.00 must include payment.* (California, Washington, D.C., New Jersey, and New York residents please include appropriate sales tax.) For billed orders, cost per copy is $7.95 plus postage and handling. (Prices subject to change without notice.)

To ensure correct and prompt delivery, all orders must give either the *name of an individual* or an *official purchase order number.* Please submit your order as follows:

Subscriptions: specify series and subscription year.
Single Copies: specify sourcebook code and issue number (such as, PE8).

Mail orders for United States and Possessions, Latin America, Canada, Japan, Australia, and New Zealand to:
Jossey-Bass Inc., Publishers
433 California Street
San Francisco, California 94104

Mail orders for all other parts of the world to:
Jossey-Bass Limited
28 Banner Street
London EC1Y 8QE

New Directions for Program Evaluation Series
Scarvia B. Anderson, *Editor-in-Chief*

Contents

Editor's Notes

The literature on applications of time series analysis to program evaluation grew rapidly in the 1960s and 1970s, in large measure in response to the pioneering analyses of designs for social experiments by Campbell (1963, 1969) and Campbell and Stanley (1963). Methodological researchers developed refined methods for overcoming threats to validity of interrupted time series experiments that were identified by Campbell and Stanley (Box and Jenkins, 1976; Box and Tiao, 1975). Textbooks and explanatory treatments — such as those of McCleary and Hay (1980); Glass, Willson, and Gottman (1975); and Cook and Campbell (1979) — have made the new methods increasingly accessible to practicing evaluators. Empirical applications are appearing increasingly often, although there is a natural lag between development of new methodology and its use in the field. Evaluation researchers must learn to use and interpret the new methods; program administrators must incorporate appropriate data collection into their programs; and sufficient time must elapse for an adequate time series to be defined. As empirical applications increase, we can expect new concepts, themes, and concerns to emerge. Application will uncover gaps between method and practice, reveal problems still to be solved, and yield new insights into the processes of social experiments.

This volume is a collection of informative applications of time series analysis to practical evaluation problems. Each of the chapters reports empirical results, discusses problems encountered in gathering and interpreting data, and addresses methodological issues that generalize to other applications. Collectively, these chapters demonstrate the range of applicability of time series analysis, reveal problems in applying the methods in real settings, and present new conceptual approaches to research and interpretation.

Applicability

In these chapters, time series analysis is applied to a wide range of problems and variables. The social programs studied and the variables used to measure impacts include the effect of family law on number of divorces (McCleary and Riggs, Chapter One); of a prison reorganization on recorded incidents and complaints (Janus, Chapter Five); of an innovative bilingual education program on achievement test scores (McConnell, Chapter Two); of a clinical intervention on productivity of retarded adults (Marsh and Shibano, Chapter Three); of a change in philosophy in a community development program on the size, number, and geographic distribution of loans (Mushkatel and Wilson, Chapter Four); and of an array of program characteristics on the amount of participation in a school lunch program (Straw, Fitzgerald, Cook, and Thomas, Chapter Six).

Time series analysis is applied in flexible combinations with other methodologies. McConnell combined time series analysis with a more traditional between-groups design to explore effects of varying amounts of bilingual education. Marsh and Shibano combined statistical with visual methods to provide practical procedures for making clinical decisions. Straw, Fitzgerald, Cook, and Thomas combined time series methods with survey methodology in order to seek out explanations for phenomena revealed by time series analysis.

Applications Problems

A useful and accurate time series is very hard to get in evaluative research. Unless the collection of appropriate data is an ongoing, high-priority part of a program's operation, data for evaluation by means of time series analysis can be inadequate or seriously flawed. It is informative to examine some of the practical dilemmas that were met and dealt with by this volume's authors. They remind evaluators again of the need for constant alertness to the mesh between program operations and evaluation.

Change in Variables. The bilingual education program evaluated by McConnell used standardized achievement tests as measures of student progress. Publishers of the tests changed editions during the course of the time series.

Short Time Series. The investigations conducted by Marsh and Shibano, by Straw, Fitzgerald, Cook, and Thomas, and by McConnell faced situations in which a program does not produce a sufficiently long series of observations for the confident application of the most effective statistical methods. When data points represent institutions or collectivities, one may continue to collect data until the number of observations is sufficient. In many cases, however, the data points represent individuals, and the logic of the clinical or educational setting precludes the arbitrary continuation of data analysis. These investigators have made contributions to methods for analyzing time series data with relatively few data points. Since these practical situations occur often, these contributions should stimulate continued research work on the analysis of short time series.

The Timing of Program Change. The study by Mushkatel and Wilson of loans made by a community development authority was motivated by an announced change in targeting philosophy. The results, however, indicated that changes in agency behavior *predated* the announced change. Apparently the concepts and initial actions involved in the change influenced the behavior of agency members before the policy change became "official." Several reanalyses were required to reveal the sequence of effects. Such situations are not uncommon in evaluative research: the *announced* onset of intervention may often be different from the *effective* onset.

Validity of Measures. Some measures of program effects, such as participation in school lunch programs, are rather direct indicators of the effects of interest. Other measures require more inference. Janus used administrative records of incidents and complaints to evaluate a change in prison management; the results were negative. But the conclusion that the program change was ineffective would require demonstration that the variables were valid measures of expectable outcomes. Janus discusses this and similar interpretive problems in detail. The validity of outcome measures is a recurring problem in evaluation research; in time series analysis, where many resources are likely to be devoted to collecting data on a few measures, the problem may be especially serious. The field is in need of strategies for incorporating construct validation — such as investigations of convergent and discriminant validity — into time series experiments.

Usefulness to Practitioners. Evaluative data should, by definition, be useful for decision making. The face validity of measures and hypotheses to decision makers is one essential factor in that usefulness. In some instances, such as instruction and clinical intervention, the decision maker must make use of data in day-to-day decisions. The decision maker cannot, or does not want to, wait for complicated statistical analyses. The graphical methods developed by Marsh and Shibano, which combine traditions of clinical decision making with statistical methods, represent an approach to meeting the needs of these practitioners. Improving accessibility of time series results to decision makers is a problem deserving increasing attention.

Conceptual Contributions

This volume is devoted to applications of time series methods, rather than with the development and presentation of new methods. But applications have a way of revealing problems and thereby leading to the development of new concepts and methods. Some new conceptual contributions have already been mentioned. Two additional contributions appear likely to have a lasting impact on evaluative reserach using time series methods.

McCleary and Riggs emphasize a concept that deserves wider attention in time series research: the construct validity of the impact model. These authors reanalyzed data for evidence of temporary and lasting effects of a change in divorce law. In the process they compare a previous analysis with the present analysis, which employs a different and arguably more appropriate model of the nature of the impact. The two analyses lead to markedly different conclusions. As the authors point out, we lack time-tested corrections for

4

threats to construct validity. The problem is similar to the construct validation of tests, and some of the approaches of that field may be helpful. Examples would include generation and comparison of multiple alternative models, and making and testing predictions based on the theoretical premises of the model. But, as McCleary and Riggs point out, there is no methodological way to assure the construct validity of the model. In effect, improving construct validity of models will require improving the theories of social process that generate the models.

Straw, Fitzgerald, Cook, and Thomas propose and illustrate a new use of time series methodology. Instead of an interrupted time series design—in which the investigators know the onset of a well-defined intervention—these investigators use time series analysis to identify trends and then to investigate potential causes of those trends. Methods for investigating causes include systematic approaches to formulating hypotheses, collecting new data by questionnaire or interview, and relating information about program changes to changes in the time series. These methods should be useful for the many evaluative situations in which there are multiple interventions not under the investigator's control and in which different units, such as schools, clinics, or administrators, can respond differently to program directors. Such development might greatly increase the usefulness of time series analysis in formative evaluation, for example, and may lead to the discovery of effective program changes that were unanticipated.

Garlie A. Forehand
Editor

References

Box, G. E. P., and Jenkins, G. M. *Time Series Analysis: Forecasting and Control.* (Rev. ed.) San Francisco: Holden-Day, 1976.

Box, G. E. P., and Tiao, G. C. "Intervention Analysis with Applications to Economic and Environmental Problems." *Journal of the American Statistical Association,* 1975, *70,* 70–79.

Campbell, D. T. "From Description to Experimentation: Interpreting Trends as Quasi-Experiments." In C. W. Harris (Ed.), *Problems of Measuring Change.* Madison: University of Wisconsin Press, 1963.

Campbell, D. T. "Reforms as Experiments." *American Psychologist,* 1969, *24,* 409–429.

Campbell, D. T., and Stanley, J. C. "Experimental and Quasi-Experimental Designs for Research on Teaching." In N. L. Gage (Ed.), *Handbook for Research on Teaching.* Chicago: Rand McNally, 1963.

Cook, T. D., and Campbell, D. T. (Eds.). *Quasi-Experimentation: Design and Analysis Issues for Field Settings.* Chicago: Rand McNally, 1979.

Glass, G. V., Willson, V. L., and Gottman, J. M. *Design and Analysis of Time Series Experiments.* Boulder: Colorado Associated University Press, 1975.

McCleary, R., and Hay, R. A., Jr. *Applied Time Series Analysis for the Social Sciences.* Beverly Hills: Sage, 1980.

Garlie A.. Forehand is a senior research scientist and director of research program planning and development at Educational Testing Service.

*A model is developed for assessing the temporary and permanent
impact of the Family Law Act, and the application and construct
validity of the model are examined.*

The 1975 Australian Family Law Act: A Model for Assessing Legal Impacts

Richard McCleary
James E. Riggs

Using the conventional notation of Campbell and Stanley (1966), the so-called time series quasi-experiment is diagrammed as

$$\ldots O \quad O \quad O \quad O \quad O\,(X)\,O \quad O \quad O \quad O \quad O \ldots$$

where each O represents an observation of a time series and where the (X) denotes an intervention presumed to have an effect on some phenomenon measured by the time series. Initially proposed by Campbell (1963), time series quasi-experiments have become widely used to measure legal impacts. Specific examples include evaluations of gun-control laws (Hay and McCleary, 1979; Deutsch and Alt, 1977; Zimring, 1975), traffic laws (Ross and others, 1970, 1982; Campbell and Ross, 1971; Glass, 1968), air pollution control laws (Box and Tiao, 1975), and divorce laws (Ozdowski and Hattie, 1981; Glass and others, 1971). Legal impact assessment relies on a contrast of pre- and post-intervention time series levels. Quasi-experimental logic cannot be divorced from the statistical models used to analyze the data, however. This is particularly true when a law has only a temporary impact on a time series.

G. Forehand (Ed.). *New Directions for Program Evaluation: Applications of Time Series Analysis to Evaluation,* no. 16. San Francisco: Jossey-Bass, December 1982.

The earliest statistical models for the time series quasi-experiment (Glass and others, 1975; Box and Tiao, 1965) assumed that impacts would be realized as abrupt, permanent shifts in the level of a series; that is

```
                              --o----o----o Postintervention
        Preintervention o----o----o--
```

Many legal impacts, of course, cannot be presumed to take this simple form. Later statistical models (Box and Tiao, 1975) relaxed the assumption of abrupt, permanent changes and took into account gradual shifts in level and temporary impact durations. One of these later model forms is the abrupt, temporary change in level; that is

```
                        --o--
                    --o--
                  --o
     Preintervention o----o----o--              Postintervention
```

This model is of great theoretical interest. In a recent evaluation of a French drinking-and-driving law, for example, Ross and others (1982) argue that the law's major impact was due not to strict enforcement but, rather, to widespread publicity. As the novelty of the new law wore off, its impact decayed. Since most theories of general deterrence highlight the role of perceptions, temporary impacts would seem to be quite common.

A practical dilemma is that permanent and temporary impacts are often statistically similar. Pre- and postintervention levels will be different in either case. Assuming a permanent impact then, the analysis may find a statistically significant permanent impact even when the "true" impact is temporary; the converse is also true. McCleary and Hay (1980) demonstrate that the static permanent impact is a special case of the dynamic temporary impact, and, relying on this formal identity, they develop a statistical method for distinguishing between permanent and temporary impacts. Permanent and temporary impacts will not be statistically distinguishable in all cases, however, so the time series quasi-experiment must rely on a theoretical specification.

This statistical dilemma is complicated further by the fact that a law may have both permanent and temporary impacts. McCleary and Hay (1980) have suggested the use of compound impact models for this purpose. In this chapter, we reanalyze an Australian divorce-rate time series using a novel compound impact model. The original analysis (Ozdowski and Hattie, 1981) assumed that reform of a divorce law would have either a permanent or a temporary impact, but not both. Our reanalysis shows that the law reform in fact had both types of impact.

The 1975 Family Law Act

The question of whether divorce-law reforms can have an impact on divorce rates has been widely debated with no clear resolution. Studies of English data by Willcox (1897) and German data by Wolf and others (1959) suggest that divorce law reforms will have only a minimal impact on divorce rates (see also, Rheinstein, 1972). The rationale of this view is that divorce rates are determined by extra-legal factors (such as economics and demographics) and, hence, that legal reforms will not have an impact on divorce rates unless the reform also has an impact on these extra-legal factors. Reanalysis of the data from Wolf and others (Glass and others, 1971) as well as a comparative study of recent divorce reforms (Eekelarr, 1978) contradict this view. Although each of these studies can be faulted methodologically, it is reasonable to assume that divorce-law reforms can have an impact on divorce rates under some circumstances, though it is not clear what these circumstances might be. Since most studies are based on reforms of the late nineteenth or early twentieth centuries, the empirical evidence, moreover, is not generalizable to modern, democratic societies.

In 1975, Australia enacted a Family Law Act, which provided for a sort of "no-fault" divorce. The act was otherwise similar to several reforms now being considered by many U.S. legislatures. Given legal, political, economic, and demographic similarities, this recent Australian experience may be the most generalizable case available.

In an evaluation of the 1975 act, Ozdowski and Hattie (1981) found only a temporary impact on divorce rates. Although the 1976 rate was nearly three times larger than the 1975 rate, the rates in 1977 and 1978 were much closer to the pre-1975 "normal" rate. Ozdowski and Hattie concluded from this evidence that "the Family Law Act has had a marked effect on the divorce trend in Australia; and the effect, however, only temporarily altered the previous 1961–1975 trend and thus one may expect the divorce rate to return to 'normal' in the near future" (p. 11). This finding supports the view that divorce-law reforms have only a minimum impact on divorce rates. Our reanalysis contradicts this finding.

The conclusion that the 1975 law had only a temporary impact is in part an artifact of data quality. The time series used by Ozdowski and Hattie ends in 1978, only three years after the 1975 act. For analytic purposes, they extended the series with a 1979 estimate of 24.6 divorces per hundred thousand population. The actual statistic for 1979 is 26.2 divorces per hundred thousand population, and, given this much higher statistic, the conclusion that divorce rates are returning to their pre-1975 level is tenuous.

Data quality notwithstanding, the statistical models used by Ozdowski and Hattie to evaluate the act cannot support conclusions about the nature of the impact. The statistical models used by Ozdowski and Hattie to analyze their time series are adequate for the limited purpose of testing a simple null hypothesis (for example, did the 1975 act have an impact on divorce rates?) but they are inadequate for the more general purpose of measuring an impact. In the next section, we review the models used by Ozdowski and Hattie and demonstrate several logical inadequacies of these models. We then develop a compound model for analyzing the impact of the 1975 Family Law Act on divorce rates. After demonstrating the use of this model in the Australian case, we propose it as a general model for the evaluation of legal impacts.

Static Impact Models

Ozdowski and Hattie (1981) based their impact analysis on separate trend lines fit to the pre- and post-1975 series. This "broken" trend-line model and the Australian divorce-rate time series are shown in Figure 1. The model may be written as

$$D_t = b_0 + b_1 t + b_2 I_t + b_3 t I_t + N_t \qquad (1)$$

where D_t is the tth observation of the divorce time series
   t is a "year counter," $t = 1946, 1947, \ldots, 1979$
   I_t is a "dummy variable," defined such that
      $I_t = 0$ until 1975
         $= 1$ thereafter
   N_t is an autoregressive integrated moving average (ARIMA) noise component

Given these definitions, the b_i parameters of model (1) have relatively simple interpretations.

   b_0 is the pre-1975 level of the D_t time series
   b_1 is the pre-1975 slope of the D_t time series
   b_2 is the post-1975 change in level of the D_t time series
   b_3 is the post-1975 change in slope of the D_t time series

The null hypothesis of no post-1975 change in the time series, $H_0: b_2 = b_3 = 0$, may be tested with t-statistics estimated for b_2 and b_3.

The impact determined by model (1), as shown in Figure 1, consists of distinct pre- and postintervention linear trends. Although this model fits the D_t time series adequately during this period, it misrepresents the underlying stochastic process. To illustrate this misrepresentation, note that model (1) will not support extrapolations of the series into the future. Extrapolation of the postintervention linear trend implies, for example, that the D_t series will take on negative values during the 1980s, which, of course, is impossible.

Figure 1. Broken Trend Line Model of Australian Divorce Rates

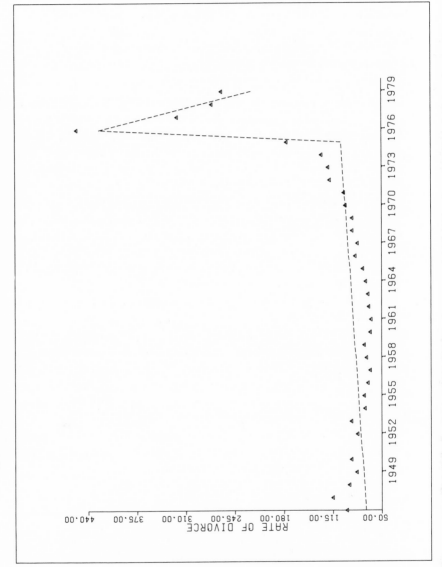

Recognizing this shortcoming of model (1), Ozdowski and Hattie built a quadratic trend model of D_t. Figure 2 shows the Australian divorce-rate time series and the quadratic trend line fit to the series.

$$D_t = 7.19 - .29T + .057T^2 + N_t \qquad (2)$$

where D_t is the tth observation of the divorce rate time series

T counts the years since 1960

N_t is stochastic component

The rationale here is that the quadratic trend model (2) describes "what would have happened" to the D_t series if the 1975 act had not been passed. Comparing the observed and forecasted values of D_t for 1979, Ozdowski and Hattie concluded that the series had returned to its "normal" level: "The predicted divorce rate for 1979 is 22.26 which is very similar to the actual rate of 24.60. It should be noted that no difference to the results above were found if the actual divorce numbers were used" (p. 11). In fact, the forecasted and observed statistics for 1979 *are* different. In absolute terms, the difference amounts to 3088 divorces but, of course, there are no criteria for determining whether a forecast error of this size is "acceptable" or "reasonable."

The forecasting model (2) has a questionable basis in any event. Like the intervention model (1), it may fit the D_t time series for a limited period; but it assumes that the stochastic process underlying D_t follows a quadratic trend. The hazards of relying on such models have been widely noted (Box and Jenkins, 1976, p. 300; McCleary and Hay, 1980, p. 35) and, in this particular case, the quadratic trend assumption is wholly unwarranted. On the basis of model (2), the D_t series will rise to 35.57 divorces per hundred thousand population by 1985. By the end of the century, the divorce rate will be 86.79 per hundred thousand population, a 700 percent increase over 1975. These forecasts are unreasonable because they imply "explosive" growth in the D_t series. And because these forecasts are unreasonable, it is unreasonable to use model (2) as a baseline measure, other things being equal, for deciding whether the impact is permanent or temporary.

Compound Impact Models

Figure 3 shows a more realistic model fit to the D_t time series. This model is represented by the difference equation

$$D_t = ((1 - B)w_1(1 - dB)^{-1} + w_2)I_t + N_t \qquad (3)$$

where B is the backshift operator defined such that $B^n D_t = D_{t-n}$

The N_t component for this model was constructed through the well-known iterative identification-estimation-diagnosis strategy developed by Box and Jenkins (1976); see also McCain and McCleary, 1979; McCleary and Hay,

Figure 2. Quadratic Trend Model of Australian Divorce Rates

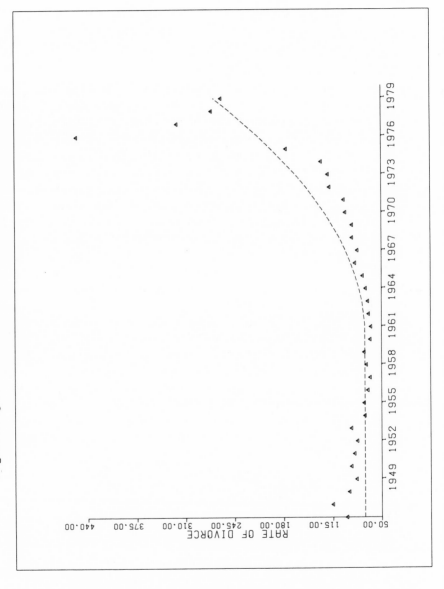

Figure 3. Compound Intervention Model of Australian Divorce Rates

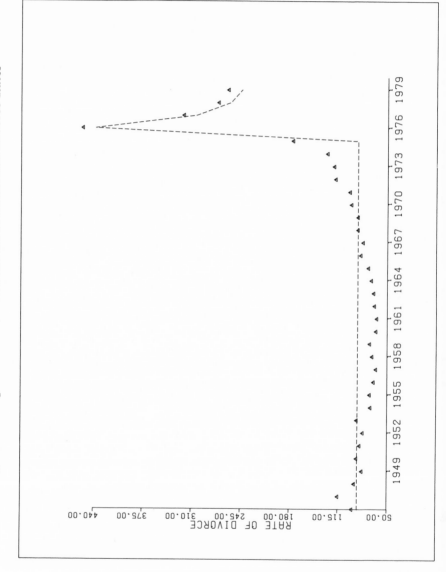

1980; McDowall and others, 1980). Since our argument deals entirely with the impact component of this model, we do not report the results of this procedure here.

We will not further develop the mathematics of this model but, instead, direct the reader to Box and Tiao (1975) and McCleary and Hay (1980, p. 187). In substantive terms, however, model (3) is the sum of an abrupt, permanent change in series level,

```
                              --o----o----o Postintervention
        Preintervention o----o----o--
```

and an abrupt, temporary (decaying) change in level,

```
                    --o--
                         --o--
                              --o
        Preintervention o----o----o--            Postintervention
```

The parameter w_1 of model (3) determines an abrupt increase in the level of the D_t series, which decays geometrically at the rate d; w_2 determines the change in level of D_t from pre- to postintervention. More specifically, in the nth postintervention observation, the level of D_t is expected to be

$$E(D_t) = d^{n-1}w_1 + w_2$$

Parameters w_1 and w_2 are scalars, but the d parameter is constrained to the "bounds of system stability,"

$$0 < d < 1$$

Given these bounds, as the postintervention series grows longer, the level of the D_t series approaches the limit

$$\lim_{n \to \infty} (d^{n-1}w_1 + w_2) = w_2$$

The expected impact of model (3) is a postintervention "spike," which decays to the preintervention level plus the quantity w_2, so model (3) permits a simple test for temporary and permanent impacts. If the estimate of w_2 is not statistically significant, we conclude that the D_t time series has decayed to its preintervention "normal" level and, thus, that the impact was only temporary.

Using an appropriate nonlinear software package (Pack, 1977), we have estimated the parameters of model (3) for the D_t time series as

$$d = .3333 \text{ with } SE = .07907; \; t = 4.21$$
$$w_1 = 20.13 \text{ with } SE = 2.4760; \; t = 8.16$$
$$w_2 = 7.769 \text{ with } SE = 2.7533; \; t = 2.82$$

All parameter estimates are statistically significant. With respect to the question of a permanent impact, we note that the t-statistic for w_2 is approximately

2.82, so the null hypothesis is rejected. The 1975 Family Law Act had a permanent impact on the D_t series as well as a short-term impact.

Using these parameter estimates and the difference equation model (3), the impact on the D_t series in 1975 and subsequent years is

$$
\begin{aligned}
d^0 w_1 + w_2 &= (.3333)^0 20.13 + 7.769 \\
&= 20.13 + 7.769 &= 27.899 \\
d^1 w_1 + w_2 &= (.3333)^1 20.13 + 7.769 \\
&= 6.709 + 7.769 &= 14.478 \\
d^2 w_1 + w_2 &= (.3333)^2 20.13 + 7.769 \\
&= 2.236 + 7.769 &= 10.005 \\
d^3 w_1 + w_2 &= (.3333)^3 20.13 + 7.769 \\
&= .745 + 7.769 &= 8.514 \\
d^4 w_1 + w_2 &= (.3333)^4 20.13 + 7.769 \\
&= .248 + 7.769 &= 8.017
\end{aligned}
$$

The temporary impact decays rapidly. By the end of 1979, the level of the D_t series is within 10 percent of its asymptotic "steady state." The net impact of this temporary effect is given by the infinite series

$$
\sum_0^\infty d^k w_1 = \frac{w_1}{1 - d} = 30.02
$$

This sum is interpreted geometrically as the area under the decaying spike in Figure 3. Given a population of 14.5 million, this temporary impact amounts to a net increase of 4,352 divorces. The permanent impact (7,769 divorces per hundred thousand population) amounts to an increase of 1,126 divorces per year in 1976 and subsequent years.

Conclusion

The 1975 Family Law Act had both temporary and permanent impacts on divorce rates. The temporary impact, due presumably to a "pent-up demand" for divorce, was so relatively large that it masked a substantial permanent impact. Our findings have specific implications for demographers. On the basis of this most recent case, it appears that divorce-law reforms can indeed have substantial impacts on divorce rates. The opposite view would seem to be ruled out by this case.

But our findings have broader implications for the analysis of time series quasi-experiments. The Ozdowski and Hattie (1981) analyses were flawed by a threat to construct validity: an inappropriate impact model. Relatively little attention has been paid to the construct validity of time series analyses and, in our opinion, these threats are always potent. Construct validity is a confusing issue in this context because its threats are poorly defined.

Threats to internal and external validity, in contrast, are finite and well defined; they may be ruled out by design. Threats to statistical conclusion validity may similarly be ruled out by ARIMA models. Threats to construct validity cannot be ruled out in this same absolute sense, however. At best, the analyst can rely on a priori theory to select a likely set of impact models. Construct validity will be only as strong as its theoretical base, and, in all cases, it will be an unknown factor.

On theoretical grounds, the impact of the 1975 Australian Family Law Act seems to be "typical." When new laws are enacted, we expect temporary effects regardless of whether or not the intervention had a "real" permanent effect, and our models should reflect this expectation. Failure to explicitly model these effects may have two consequences. First, a "demonstration" or "placebo" effect may be mistaken for a permanent "real" effect. Second, as demonstrated in this case, a permanent "real" effect may be hidden by a temporary reactive effect. In either case, failure to control these confounding effects may have disastrous consequences.

References

Box, G. E. P., and Jenkins, G. M. *Time Series Analysis: Forecasting and Control.* (Rev. ed.) San Francisco: Holden-Day, 1976.

Box, G. E. P., and Tiao, G. C. "A Change in Level of a Nonstationary Time Series." *Biometrika,* 1965, *52,* 181-192.

Box, G. E. P., and Tiao, G. C. "Intervention Analysis with Applications to Economic and Environmental Problems." *Journal of the American Statistical Association,* 1975, *70,* 70-79.

Campbell, D. T. "From Description to Experimentation: Interpreting Trends as Quasi-Experiments." In C. W. Harris (Ed.), *Problems of Measuring Change.* Madison: University of Wisconsin Press, 1963.

Campbell, D. T., and Ross, H. L. "The Connecticut Crackdown on Speeding: Time Series Data in Quasi-Experimental Analysis." *Law and Society Review,* 1971, *3,* 33-53.

Campbell, D. T., and Stanley, J. C. *Experimental and Quasi-Experimental Designs for Research.* Chicago: Rand McNally, 1966.

Deutsch, S. J., and Alt, F. B. "The Effect of Massachusetts' Gun Control Law on Gun-Related Crimes in the City of Boston." *Evaluation Quarterly,* 1977, *1,* 543-568.

Eekelarr, J. *Family Law and Social Policy.* London: Weidenfeld and Nicolson, 1978.

Glass, G. V. "Analysis of Data on the Connecticut Speeding Crackdown as a Time Series Quasi-Experiment." *Law and Society Review,* 1968, *3,* 55-76.

Glass, G. V., Tiao, G. C., and Maguire, T. O. "The 1900 Revision of German Divorce Laws." *Law and Society Review,* 1971, *8,* 539-562.

Glass, G. V., Willson, V. L., and Gottman, J. M. *Design and Analysis of Time Series Experiments.* Boulder: Colorado Associated University Press, 1975.

Hay, R. A., Jr., and McCleary, R. "Box-Tiao Time Series Models for Impact Assessment: A Comment on the Recent Work of Deutsch and Alt." *Evaluation Quarterly,* 1979, *3,* 277-314.

McCain, L. J., and McCleary, R. "The Statistical Analysis of the Simple Interrupted Time-Series Quasi-Experiment." In T. D. Cook and D. T. Campbell (Eds.), *Quasi-Experimentation: Design and Analysis Issues for Field Settings.* Chicago: Rand McNally, 1979.

18

McCleary, R., and Hay, R. A., Jr. *Applied Time Series Analysis for the Social Sciences.* Beverly Hills: Sage, 1980.

McDowall, D., McCleary, R., Meidinger, E. E., and Hay, R. A., Jr. *Interrupted Time Series Analysis.* Beverly Hills: Sage, 1980.

Ozdowski, S. A., and Hattie, J. "The Impact of Divorce Laws on Divorce Rate in Australia: A Time Series Analysis." *Australian Journal of Social Issues,* 1981, *16,* 3-17.

Pack, D. J. *A Computer Program for Analysis of Time Series Models Using the Box-Jenkins Philosophy.* Hatboro, Pa.: Automatic Forecasting Service, 1977.

Rheinstein, M. *Marriage, Stability, Divorce and the Law.* Chicago: University of Chicago Press, 1972.

Ross, H. L., Campbell, D. T., and Glass, G. V. "Determining the Effects of a Legal Reform: The British 'Breathalyzer' Crackdown of 1967." *American Behavioral Scientist,* 1970, *13,* 493-509.

Ross, H. L., McCleary, R., and Epperlein, T. "Deterrence of Drinking and Driving: An Evaluation of the French Law of July 12, 1978." *Law and Society Review,* 1982, *16,* 2.

Willcox, W. F. *The Divorce Problem: A Study in Statistics.* New York: Columbia University Press, 1897.

Wolf, E., Lüke, G., and Hax, H. *Scheidung und Scheidungsrecht: Grundfrage der Ehescheidung in Deutschland.* [Divorce and Divorce Laws: Hypothesis About Divorce in Germany]. Tübingen: J. C. B. Mohr, 1959.

Zimring, F. "Firearms and Federal Law: The Gun Control Act of 1968." *Journal of Legal Studies,* 1975, *4,* 133-198.

Richard McCleary is associate professor of criminal justice at the State University of New York, Albany.

James E. Riggs is senior research analyst with the Arizona Department of Corrections.

A multiyear evaluation of the long-range benefits of bilingual education illustrates the utility of a time series design for programs involving minority populations.

Evaluating Bilingual Education Using a Time Series Design

Beverly B. McConnell

Researchers in the past two decades have shown an increased interest in the application of time series designs in evaluation. Much of this application has been in single-subject research where either the unique characteristics of the subject or the treatment approach does not lend itself to group designs. There has also been a growing application of time series designs in analyses of historical or natural phenomena and social reform in which the nature of the variables studied is such that they either cannot be manipulated or reduced to laboratory conditions (Glass and others, 1975).

Several authors have proposed that it would be useful to combine time series designs with more conventional designs. (See discussion in Kratochwill, 1978, p. 79). Research or evaluation studies using such a combined approach are still relatively rare. The study presented in this chapter, however, represents such a combined approach.

Why New Evaluation Methods Are Needed in Evaluating Bilingual Education

By definition, bilingual education programs, such as the one used in this study, involve linguistic and ethnic minorities. This fact severely restricts

G. Forehand (Ed.). *New Directions for Program Evaluation: Applications of Time Series Analysis to Evaluation,* no. 16. San Francisco: Jossey-Bass, December 1982.

the utility of most of the traditional designs used in educational evaluation. From a technical standpoint, evaluation designs involving random assignment would be the preferred method. However, random assignment of children to groups, particularly if this includes a "no treatment" group, is considered a violation of their civil rights. This position is reinforced by several court decisions at both the state and federal levels (Applewhite, 1979).

Use of a comparison group design, with nonrandomly assigned groups, seldom produces groups that are truly comparable on the key factor of initial linguistic competence. There is a good reason for this. Administrative regulations by agencies funding bilingual education stipulate that the children enrolled must be those "most in need." This places the children with the most restricted English skills into the bilingual program, and leaves for a comparison group children who may represent the same ethnic group but who speak primarily English or at least have less of a problem with English than those selected for bilingual education.

The purpose of the comparison group is to provide a "reasonable basis" for judging how children would have done without a special program. All too often evaluators of bilingual programs have assumed that if the children in bilingual and comparison classrooms both represent the same ethnic background or socioeconomic status a good match has been made. If the comparison group children, however, speak only or primarily English there clearly is no "reasonable basis" for assuming that their progress in all-English classrooms represents the same kind of progress that would be made by a non-English-speaking child. The only national study that has been conducted of bilingual education, commonly referred to as the AIR Report, represents just such a mismatch in a comparison group design. Out of the Hispanic comparison group selected, it was found that 83 percent spoke only English. It is therefore probably not surprising that their pre- to posttest gains in English language arts were somewhat better than those of children in the bilingual programs, only 26 percent of whom were judged to be English speaking, with 74 percent classified as bilingual or monolingual in Spanish (American Institutes for Research, 1977, Appendix pp. 123–127).

This is not an isolated example. A review of other evaluation studies of bilingual education in the United States indicates that selection factors resulting in great differences in comparative proficiency in English and another language have presented a problem in the majority of studies of bilingual education using a comparison group design (Baker and de Kanter, 1981).

The most frequently used evaluation designs for education, however, are a group of designs referred to collectively as "norm referenced" designs. There are a variety of these, which differ in the statistical treatment applied, but all of them make use of standardized tests for which there are published norms, usually derived from some national sampling of United States school children. Herein lies the problem for use of this type of evaluation with ethnic and linguistic minorities. Proportionately, these minorities are not much represented in the samples used to determine the published norms of the tests. This means that basing an expectation of a "no treatment effect" on the performance of children in the test norm group will seriously bias the results to the extent that the test relies on English language or knowledge relating to cultural experience that the majority group child may have access to but the minority child does not (see Coleman and others, 1966; Linn, 1979; Mercer, 1973).

It is not the purpose of this chapter to elaborate on the bias resulting from the application of inappropriate test norms to minority populations except to make the point that this type of evaluation design and most of the other traditional approaches to evaluation have serious defects when applied to students within linguistic and ethnic minorities. There is, therefore, a great need for innovation in evaluation approaches for these groups, and the use of a time series design merits careful attention in these circumstances.

The Advantages of Time Series Designs
for Evaluating Minorities

When a "unit repetitive" time series design is used, children are used as their own controls. This eliminates most of the selection problems that have plagued comparative group research described above. Even when the analysis involves observations of different children, a "unit replicative" approach to use Glass's terminology (Glass and others, 1975), the same selection criteria are used for successive groups of children who replicate the treatment. There is a much greater chance of finding a similarity in children who have been selected "in" to a bilingual program, than finding such a similarity between groups in which key variables have been used to select some children "in" and other children "out" of an educational treatment such as bilingual education. Using stratified analysis, it will probably be possible to secure a fairly good match on ethnic and language competence variables in subsamples of children within successive waves of children served by a bilingual program.

Since the time series design does not require withholding treatment from some group of children for purposes of evaluation design, it has obvious merits when there are legal implications involved, as is true in this case.

Time is also potentially quite useful as a dimension in the evaluation of programs such as bilingual education where short- and long-term results may be contradictory. This occurs because children learning in two languages usually sequence the instruction so that at certain points in their education, when they have not yet received instruction in the "other" language, there will be a disadvantage, which then disappears within the long-range evaluation of program results (see Rosier and Holm, 1980).

The problem of tests that represent a cultural bias remains, but it can be largely neutralized if all children with whom other children are being compared represent the same ethnic and language background. Whatever bias exists will affect all groups equally and the patterns of relationships will represent an internal validity.

Description of the Experimental Program

The educational program reported here is a demonstration program known as IBI (Individualized Bilingual Instruction) that was started in 1971 and is now in its eleventh year of operation. The program serves children of preschool age and kindergarten through third grade. Because is was planned as a national demonstration program for children of migrant farm workers, it operates parallel programs in a number of different locations. One program is located in south Texas in a "home base" area, and others are located twenty-five hundred miles away in some small towns in Washington state where Texas migrants move during the season. With a good deal of seasonal fluctuation, the program operates most of the year in both states because the migrants come and go at different times, and some move around from one employer to another, all within the same area.

The program also operates a "mobile component." When the families leave Texas for work areas in the north, the teachers leave too (since the teaching staff are adults recruited from migrant families), and temporary programs are set up with the cooperation of educational agencies in the "receiving states." The educational program for school-age children requires about one hour a day, and is usually worked out on a released-time basis from other classes. Preschool and kindergarten children have about two hours of instruction, with structured lessons alternated with informal free choice activities appropriate at this age level. The mobile program has set up temporary programs in Illinois, Idaho, Oregon, and Washington states. At the end of the work season, the families return to Texas, and the educational program continues there during the winter months.

More detailed descriptions of the experimental program are available from other sources (McConnell, 1981a, 1981b). Teachers in the program are bilingual, and IBI uses an individualized curriculum designed to teach oral proficiency and literacy in both Spanish and English, as well as math and handwriting.

How the Evaluation Model Was Developed

Because of the irregular attendance patterns characteristic of migrant children, most attempts at testing all children on the same calendar dates in fall and spring will prove futile. There is a great likelihood that the majority of the children present at one time period will not be present at the next one. In addition, the posttest in the spring may be measuring the effects of schooling lasting only a few weeks for some children and for several months for others. From the outset, therefore, the IBI program followed an individualized testing schedule based on the attendance of an individual child. Pretests were completed when the child enrolled and after each hundred days of program attendance. If the child attended for sixty-six days and then left for seven months and returned, the first day back became day sixty-seven. A child who attended regularly might complete one hundred days attendance in five months. For another child, it might take over a year.

Each month the testers were notified at each site which children had reached an attendance interval requiring another series of tests and these tests were given. New children were allowed to enroll in the program in any month throughout the year when the program was operating, so pretests were also being given throughout the year. All of the tests selected were tests that do not require administration during restricted time periods for the interpretation of test scores, but that have standard score conversions tied to a child's age rather than grade level and that can be given at any time.

In the first few years, the program evaluated child gain scores against arbitrarily set program "goals." Children's tests were grouped according to their primary language classification and age at time of testing, and their performance was judged by whether they had met the goals set for different periods of attendance. This had the merit of representing a similar amount of program intervention for the children whose tests were grouped together for analysis. It had the drawback of providing no means of judging how the children might have done without the intervention.

In the fourth year of program operation, the evaluation design was changed. The accumulation of pretest scores of children since the program began operating was used to represent how the children would have fared in traditional education available to them, excluding the influence of the IBI supplemental program. For each individual child this pretest score represented the effect of whatever collection of unrelated and possibly contradictory approaches to education the migrant child might have encountered in his moves from school to school. It seemed reasonable to assume that collectively these scores would represent the effects of "traditional" education for migrant children. The posttest scores of children who had attended the bilingual program for varying periods of time were then compared statistically to the accumulated pretest scores of other children of the same age and language classification who were just starting the program. This analysis was used to see if there was a statistical superiority for children who had received the bilingual program, and whether longer periods of attendance produced continued increases in test scores.

Since children were allowed to enter the program at any age between age three and third grade, pretests were available for every age level. (The program only served children through the third grade.) This meant that at age six, for example, the program had pretest scores of children who started the bilingual program at that age. It had posttest scores of other six-year-old children who had enrolled at an earlier age and who had attended the bilingual program for one, two, or even three school years by age six. For the annual program evaluation (McConnell, 1981a), analysis was done separately for each age level based on subgroupings by period of attendance and children's primary language classification. This represented a time series design in that each child received a series of observations after different periods of intervention. By incorporating a cross-sectional analysis of children tested at different points in the time interval series for a given evaluation, it follows a conventional evaluation design for analysis of variance between groups.

Using the notational system developed in the major discussion of time series analysis by Glass and others (1975), the evaluation design would look as shown in Figure 1.

Each row in Figure 1 represents individual children who have entered the program at a different time and different age levels. The children have been pretested (0_1) and then received continuous intervention (I) and have been posttested after given intervals of attendance $(0_2 \ldots$ and so on). The vertical lines represent an analysis unit of the tests used for a given age level or for a given time period. Since children will have received the intervention for different periods of time within the analysis unit, the effects of the program can be measured against time of attendance as the key variable, holding constant age group, year of testing, or whatever other variables are desired in a particular analysis.

For the purpose of describing the method for this chapter, a special analysis was run on test data for children tested over a six-year period, 1976 through 1981. Standard scores are used so the analysis could combine tests of children across age groups. On one test, standard scores are available only for children over the age of five, so this analysis is limited to children over the age of five (although the program accepts children beginning at age three). It is also limited to children whose primary language is Spanish. And to simplify the visual presentation of data, three sets of test years have been used combining 1976–1977, 1978–1979, and 1980–1981. Likewise, tests after intervals of one hundred and two hundred days have been combined into a category considered the equivalent of one one-hundred-eighty-day school year; three-hundred- and four-hundred-day tests are combined for the two school-year category; five-hundred- and six-hundred-day tests are combined for a three school-year category. Any tests beyond six hundred days have not been used in the analysis since there were not enough of these in each set of test years to provide stability of statistical analysis.

Figure 1. IBI Evaluation Design

```
 O₁    I O₂    I O₃    I O₄    I O₅    I O₆  │ I O₇ │ I O₈
----------------------------------------------┤      ├----------
        O₁     I O₂    I O₃    I O₄    I O₅  │ I O₆ │ I O₇
        ----------------------------------------┤      ├----------
               O₁     I O₂    I O₃    I O₄  │ I O₅ │ I O₆
               ----------------------------------┤      ├----------
                      O₁     I O₂    I O₃  │ I O₄ │ I O₅
                      ----------------------┤      ├----------
                             O₁     I O₂  │ I O₃ │ I O₄
                             --------------┤      ├----------
                                    O₁    │ I O₂ │ I O₃
                                    ------┤      ├----------
                                          │  O₁  │ I O₂
                                          └------┤----------
```

0 = Observations for individual children numbered serially

I = Intervention, i.e., 100 days attendance in bilingual program

UNITS
OF
ANALYSIS
FOR A
GIVEN
EVALUATION

The tests on which this analysis is based are the Peabody Picture Vocabulary Test (Dunn, 1965) in English, with an alternate form adapted to measure Spanish vocabulary, and the Wide Range Achievement Test (Jastak and Jastak, 1965) including subtests for English reading and for arithmetic. Although the publisher of each of these tests has brought out a revision in the last few years, the older forms of the tests have been retained for the sake of acquiring longitudinal data without changes of instrumentation.

The Analysis of IBI Program Results

In most time series evaluations the baseline data represent several observations before an intervention is started. The purpose of repeated measures is to establish that there is a pattern to the observations so that it will be possible to judge whether or not the intervention has, in fact, changed the pattern. In this case there was only one observation for each child, the pretest, before intervention was started. However, the pattern of the baseline data is very clear from the successive pretests of children entering the program over a six-year period. As shown in Figures 2 and 3, the observations at "0 Attendance" represent a very low level of academic and language proficiency. Analysis of variance of just the pretests in each of the three sets of test years shows that there is no significant difference in this pattern of entry-level scores over the six-year period. This uniformity of entry-level scores applies to all four subjects (arithmetic, reading, English, and Spanish) tested.

Figure 2 shows the mean standard scores for children tested after attendance periods which are the equivalent of one, two, and three school years. This data is presented separately for three successive two-year time periods, which is the equivalent of showing three successive replications of the experimental evidence. Once again a very similar pattern is revealed. Each period of attendance in the bilingual program produces a marked gain in academic skills.

The figures also show the cumulative mean for all six years at the different attendance levels. In arithmetic, the mean scores of children entering the program were below that of all but 18 percent of U.S. school children. After one year, the mean scores were in a low average range. After two years, the mean was at national norms, and at three years, above the average score of two-thirds of U.S. school children.

In English reading, the gains also show an increase for each time period. Initial scores are very low, the mean pretest score being below the twelfth percentile. After two years attendance, the mean score has increased enough to be within the low average range relative to the native English-speaking children whose scores were used to determine the norm. By three years, the English reading scores are approaching national norms, with a mean above the fortieth percentile.

Figure 2. Mean Standard Scores by Years of Attendance
on Wide Range Achievement Test

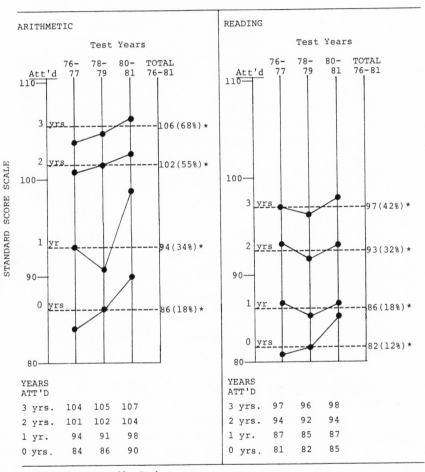

*National Percentile Rank

Over the six-year period the same pattern of academic gains can be observed with successive waves of children going through the program, indicating that the program effect is achieved despite teacher turnover and the natural variance of ability among children. Analysis of variance among the three sets of test years yielded no significant differences in the pattern of scores on any of the tests except arithmetic, where the higher scores in 1980–1981 represented a significant difference from the other two sets of data. However, the pattern of increased achievement based on longer periods of attendance was the same. At the same time, analysis of variance based on period of program attendance was highly significant.

To summarize, the findings presented in Figure 2 demonstrate that with a program of bilingual education, the Spanish-speaking sons and daughters of migrant farm workers are able to bring their academic skills up to a level comparable to native English-speaking U.S. school children in approximately a three-year period. This finding has been replicated a number of times with successive waves of children, giving added confidence that this represents a stable program effect.

Figure 3 illustrates the changing pattern of English and Spanish vocabulary scores with successive waves of children. As demonstrated in this figure, there was initially a very great disparity between children's vocabulary in Spanish, their primary language, and in English. Each year of program attendance produced a large gain in English vocabulary by children enrolled in the bilingual program. The cumulative mean score over all six years of test data shows that children who attended the bilingual program for three years had a mean score of 72 in English vocabulary and 75 in Spanish vocabulary. It would appear that within a three-year period children have reached a balanced bilingualism in the language skills measured by these tests. They have improved their Spanish at the same time that they have greatly increased their ability in English, which is their second language.

In Spanish vocabulary there is a more complicated pattern than in the other subjects where there was a straight line correlation between length of time in the program and higher scores. Children's mean scores in Spanish showed small but statistically significant gains after one and two years in the bilingual program. In the third year the mean score dropped slightly. This pattern was true for test scores during 1976–1977, in 1978–1979, and in 1980–1981. The third year drop in mean score is small enough that it might be attributed to chance variation or children reaching an apparent ceiling on this test. However, when the pattern is so consistently repeated, it appears more likely that it represents an unexpected outcome of the program. The third year of program attendance is the time when children's English vocabulary scores have become nearly as high as their Spanish scores. It may be that the greater utility of English in the social and economic setting of the United

Figure 3. Mean Standard Scores by Years of Attendance
on Peabody Picture Vocabulary Test

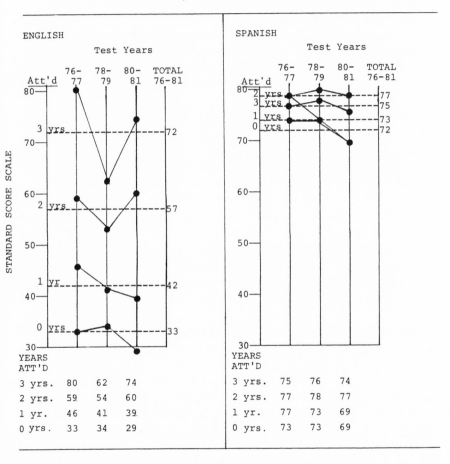

States causes children to make a language shift as soon as their English has developed to a sufficient point to make this feasible. This appears to have happened at some cost to the continued development of their Spanish. This is the type of finding that would not be revealed except through the use of time series analysis.

How Does the IBI Evaluation Meet Tests of Internal Validity?

Campbell and Stanley (1966), in their now classic work on experimental and quasi-experimental designs, identify a number of factors that might provide alternative explanations for experimental results. They called these "threats" to internal or external validity. Glass and others (1975) added some additional "threats to validity" unique to time series designs. Since the evaluation described in this chapter is presented as a possible prototype of a new approach to evaluating educational programs, particularly those involving linguistic minorities, it may be useful to discuss some of these "threats to validity."

History. Time series designs are particularly susceptible to the possibility that some external event may have caused the observed changes. However, this appears an unlikely explanation in this case. The fact that there has been no significant change in the level of skills for new children entering the program over a six-year time period would seem to indicate that there has not been any generalized improvement in the educational prospects of migrant children in the general school system that could account for the changes reported in the evaluation.

Maturation. Use of standard scores in this analysis factors out the effects of maturation since standard scores are age adjusted. However, there is always some distortion in using standard scores with a population quite different from the experimental group. If the norms do not fit, the growth curve adjusting for age differences will not fit either. To get away from this source of distortion, the usual evaluation pattern followed in the IBI program, when a more detailed presentation of findings is possible, is to use raw scores for analysis and to present the findings separately for each age level (see McConnell, 1981a). The basic findings, however, are the same as have been presented in this chapter. Within every age level there is a consistent and significant increase in test scores based on children's length of attendance in the bilingual program.

Testing and Instrumentation. The same test instruments have been retained throughout the test period. When children are retested on the same instruments, there is some possibility of change due to "learning the test." However, most test learning has been found to occur between the first and second administration of the test. The continued gains over a period of years in this case make it unlikely that test learning has contributed much to the results.

Hawthorne Effects. It seems unlikely that the special effects caused by extra attention to an experimental group could be sustained through as many years as this study represents.

Reactive Intervention. This factor, identified by Glass and others (1975), refers to experiments started at an artificially low point so that change represents a return to "normal" conditions. Its counterpart is the regression effect in test scores. Regression effects undoubtedly do explain part of the increase in scores within the IBI program because of the extremely low level of scores of children who entered the program, particularly in English vocabulary. Regression obviously cannot be the explanation of all of a change of the magnitude reported in this paper.

Summary and Conclusion

The method presented in this chapter combined time series analysis techniques with the more conventional designs based on variance between groups. Based on the collection of nearly nine hundred observations on each of four tests over a six-year period, it demonstrates that Spanish-speaking children who are initially very low in English skills can be greatly benefited by a program of bilingual education such as IBI. Children can be brought up to a level of balanced bilingualism in approximately a three-year period of intervention, and in this time they will have improved their Spanish as well. It also demonstrates that children who are extremely weak in academic skills can, through this type of bilingual program, reach a level that is competitive with mainstream, native English-speaking school children.

The limitations of some of the other methods of evaluation now in use to measure the effectiveness of education programs for linguistic minority children have been discussed in this chapter. The method presented here could be adapted to most bilingual settings and appears to have much to offer in the special circumstances surrounding the evaluation of educational gains by linguistic and ethnic minority children.

References

American Institutes for Research. *Evaluation of the Impact of ESEA Title VII Spanish/ English Bilingual Education Program.* Vol. I: *Study Design and Interim Findings.* Palo Alto, Calif.: American Institutes for Reserach, 1977.

Applewhite, S. R. "The Legal Dialect of Bilingual Education." In R. V. Padilla (Ed.), *Ethnoperspectives in Bilingual Education Research.* Vol. 1: *Bilingual Education and Public Policy in the United States.* Ypsilanti, Mich.: Eastern Michigan University, 1979.

32

Baker, K., and de Kanter, A. A. "Effectiveness of Bilingual Education: A Review of the Literature." Paper prepared for the Office of Planning, Budget, and Evaluation. Washington, D.C., 1981.

Campbell, D. T., and Stanley, J. C. *Experimental and Quasi-Experimental Designs for Research.* Chicago: Rand McNally, 1966.

Coleman, J. S., Campbell, E. Q., Hobson, C. J., McPartland, J., Mood, A. M., Weinfeld, F. D., and York, R. L. *Equality of Educational Opportunity.* Washington, D.C.: U.S. Government Printing Office, 1966.

Dunn, L. M. *Peabody Picture Vocabulary Test.* Circle Pines, Minn.: American Guidance Service, 1965.

Glass, G. V., Willson, V. L., and Gottman, J. M. *Design and Analysis of Time Series Experiments.* Boulder: Colorado Associated University Press, 1975.

Jastak, J. F., and Jastak, S. R. *The Wide Range Achievement Test.* Wilmington, Del.: Guidance Associates, 1965.

Kratochwill, T. R. *Single Subject Research: Strategies for Evaluating Change.* New York: Harcourt Brace Jovanovich, 1978.

Linn, R. L. "Validity of Inferences Based on the Proposed Title I Evaluation Models." *Educational Evaluation and Policy Analysis,* 1979, *1* (2), 23–32.

McConnell, B. B. *Long Term Effects of Bilingual Education.* Pullman, Wash.: Bilingual Mini Schools, 1981a. ERIC Document Reproduction Service No. ED 206 203.

McConnell, B. B. "Plenty of Bilingual Teachers." In R. V. Padilla (Ed.), *Ethnoperspectives in Bilingual Education Research.* Vol. 3: *Bilingual Education Technology.* Ypsilanti, Mich.: Eastern Michigan University, 1981b.

Mercer, J. P. *Labeling the Mentally Retarded.* Berkeley: University of California Press, 1973.

Rosier, P., and Holm, W. *The Rock Point Experience: A Longitudinal Study of a Navajo School Program (Saad Naaki Bee Na'nitin).* Bilingual Education Series No. 8. Arlington, Va.: Center for Applied Linguistics, 1980.

Beverly B. McConnell is the evaluation director for IBI, Individualized Bilingual Instruction, a national research and demonstration program in bilingual education that has its evaluation office in Pullman, Washington.

A procedure combining visual analysis and ordinary least squares regression provides a straightforward approach for evaluating clinical interventions using time series data.

Visual and Statistical Analysis of Clinical Time Series Data

Jeanne C. Marsh
Matsujiro Shibano

In the last fifteen years, researchers have developed interrupted time series designs that offer important and more easily implemented alternatives to group designs in the evaluation of clinical interventions (Browning and Stover, 1971; Hersen and Barlow, 1976; Kratochwill, 1978). The development of these designs owes much to the conceptual work of Skinner (1953), Sidman (1960), Chassan (1967), and Campbell and Stanley (1966). For clinical evaluation, the advantages of designs based on repeated measurement over time within a single individual include the designs': (1) compatibility with the clinical practice of treating one individual at a time; (2) measurement of dependent variables most relevant to individual clients' problems; (3) measurement of dependent variables over time, allowing for the analysis of the process of change in clients' behavior; and (4) achievement of control through a baseline phase and repeated measurement over time. As a result of these advantages, time series designs are being used increasingly by clinical

This chapter is based on research supported by the Illinois Department of Mental Health and Developmental Disabilities, grant #8248-11.

G. Forehand (Ed.). *New Directions for Program Evaluation: Applications of Time Series Analysis to Evaluation,* no. 16. San Francisco: Jossey-Bass, December 1982.

researchers and practitioners to evaluate interventions. The congeniality of time series research designs with clinical practice has even led some to the prediction that these designs will reduce the "research-practitioner split" (Hersen and Barlow, 1976).

With respect to the analysis of time series data, clinical evaluators have depended almost entirely on careful visual inspection of graphed time series. Some descriptive statistics such as means, percentages, and ratios are used to facilitate visual interpretation. Inferential statistics, indispensible to the group comparison approach, have been used very little. Further, when some researchers have attempted applications of inferential statistical methods to interrupted time series data, they have been strongly criticized by their colleagues (Baer, 1977; Michael, 1974b). The criticisms are of three types: (1) inferential statistical methods depend upon the underlying assumption of independent observation; this assumption is violated when applied to time series data, thus the methods are not appropriate for analyzing individual performance data; (2) successful experimental control or manipulation should result in dramatic change that can be easily detected by visual inspection without statistical procedures; and (3) statistical procedures, especially those for sophisticated inferential statistics, are not practical in evaluating clinical situations because mastery of the procedures may require considerable mathematical training.

The relatively recent development of nonlinear regression procedures for the analysis of interrupted time series data repudiates these criticisms. Based primarily on the work of Box and Jenkins (1970) and Box and Tiao (1975), several authors have elucidated procedures for interrupted time series analyses appropriate for clinical data with twenty or more observations in each phase (Glass and others, 1975; Gottman and Glass, 1978; Gottman, 1981; Horne and others, 1982; McCain and McCleary, 1979; McCleary and Hay, 1980). These procedures address the above criticisms in three ways: (1) they do not violate the assumption of statistically independent obervations; (2) they may detect in time series data significant change that may be difficult to detect visually; and (3) they can be understood intuitively by researchers and practitioners with limited mathematical or statistical backgrounds. Each of these criticisms is discussed briefly below.

Conventional inferential statistical methods, such as the analysis of variance (ANOVA), the t-test, and regression, have been important tools for group comparison designs. However, most of these procedures can be used appropriately only when the basic assumption of statistically independent observations is satisfied. Gentile and others (1972) used analysis of variance procedures with clinical time series data by assuming that each observation in the baseline and treatment could be treated as an independent case. For criticism and discussion of this approach, see Gottman and Glass (1978), Hartmann (1974), and Michael (1974a, 1974b). In clinically relevant time series data, observations often are not statistically independent. The estimated frequency with which serial correlation of this type occurs varies; 21 percent of the series analyzed by Marsh and Shibano (1982), 29 percent of series reviewed by Kennedy as reported by Hartmann and others (1980), and 83 percent of the series reviewed by Jones and others (1977) were serially correlated. Serial dependency among observations seriously restricts the use of conventional statistical procedures. Nonlinear regression procedures take account of serial dependency and use it to characterize the nature of the series.

As mentioned previously, visual analysis currently represents the primary strategy for the analysis of clinical time series data. When visual analysis is undertaken, there are three important factors that must be examined: the variability, the level, and the trend of the series. An abrupt and large-level difference between phases (baseline and intervention) is the most convincing evidence for significant intervention effects. Extreme variability and upward or downward trends are factors that diffuse a level difference making evaluation of the treatment effect more difficult. Those who advocate solely visual analysis argue that the existence of high variability and trends is largely an indication of weak experimental control (Parsonson and Baer, 1978). It is obvious that some measurements may be highly variable despite careful experimental control. Further, in settings such as schools, homes, and communities (as compared with the laboratory), variability and trends are more likely to occur since it is harder to fully control various potential competing variables (Kazdin, 1976; Gottman and Glass, 1978). As applied researchers and practitioners work increasingly in community settings, it is important to use statistical procedures that are capable of analyzing data containing considerable variability as well as trends.

Finally, the approach described here does not require sophisticated mathematical or statistical knowledge. Advances in computer technology allow for the completion of statistical operations with procedures that are straightforward and intuitively understandable. Outputs and products are visually informative and easy to interpret.

The primary problem that remains in the application of statistical time series techniques to clinical data is the small number of data points typically available. Time series data used to analyze policy interventions, such as crime statistics, mortality rates, or economic indicators (for example, gross national product) often are collected and maintained over long periods of time. In contrast, the exigencies of clinical intervention determine the length of time during which data are collected in clinical evaluations. Decisions about introducing and withdrawing interventions are made on the basis of the pattern of the data or on the basis of ethical or treatment considerations. These decisions determine the length of the baseline and treatment phases and thus the number of observations available for analysis. Typically, data used to analyze clinical interventions contain many fewer than the twenty (Hartmann and others, 1980) to fifty (Box and Jenkins, 1970) data points per phase recommended for nonlinear regression techniques.

Short series are a problem for two reasons. Model fitting is performed with less confidence in the adequacy of the model, and statistical tests are less likely to detect true differences. Since many clinical time series contain fewer than twenty observations, particularly in the baseline phase, many clinical series are uncomfortably short for analysis with nonlinear regression methods.

The statistical approach described here is based on ordinary least squares regression and integrates visual and statistical analyses. The procedure is adapted from procedures used by business problems (Roberts and others, 1981); it has the advantages of nonlinear regression procedures described above but requires fewer observations. Fewer parameters typically are estimated within this procedure and therefore fewer data points are required. Given that the series can be described with a relatively simple model, linear regression — especially in combination with visual analysis — can be used to determine a model for short series. Once the model is selected, the intervention effects may be tested. The approach is outlined below and then elaborated in the discussion and example that follow:

Phase I
1. Determination of stationarity of series
2. Identification of an autoregressive model
3. Assessment of goodness-of-fit and adequacy of model

Phase II
1. Estimation of the impact of the intervention
2. Assessment of goodness-of-fit and adequacy of intervention impact model.

A Regression Approach to Intervention Analysis

A time series is viewed a having two components: a stochastic component determined by the influence of random events on the series and a deterministic component representing the influence of events, such as treatment, that are consistent across time. There are two types of stochastic processes. An *autoregressive* process (AR) is one in which an observation at the present time depends on previous observations. A *moving average* process (MA) is a weighted sum of previous random shocks to the series. Analysis of the series requires identification of the stochastic component followed by estimation of the deterministic component or intervention impact. The approach presented here is an application of the regression approach for analyzing the impact of an intervention on a single series of data (Roberts and others, 1981). This model is appropriate for evaluating the permanent impact of an intervention upon a stationary series, that is, a time series without trends before and after the intervention. A nonstationary series requires a preliminary trend-removing transformation; other models may be more appropriate with such series (Box and Tiao, 1975; Horne and others, 1982).

Stochastic Model Component. The first step in the analysis is the visual and statistical examination of the time series to determine whether it is stationary. The pre- and postintervention series are analyzed separately. When the series is short, some authors have suggested the possible use of both series combined for the model building (Glass and others, 1975; McCleary and Hay, 1980; Horne and others, 1982). However, combined series may lead to inadequate model identification. For example, if a level difference exists between the pre- and postintervention phases, a high autocorrelation function typically will result even when the within-phase series is random. Simply stated, combining two sequential stationary series with a level difference creates a nonstationary series. Thus, the model identification phase in this procedure relies on the preintervention series alone. Stationarity is determined by visual inspection of the plotted data and the careful examination of the autocorrelation function (ACF).

Given that the series is stationary both before and after the intervention, one may proceed to the next step. Here the focus is on determining a preliminary autoregressive model. When the underlying stochastic process of a given series can be described by a purely autoregressive model, the autocorrelation and partial autocorrelation functions (ACF and PACF) have distinctive features. When the series is a purely autoregressive process of the pth order, its ACF dies out and its PACF cuts off after the pth order.

In the next step of the analysis, the adequacy of the identified model is assessed by implementing the ordinary least squares regression with the appropriate number of previously identified autoregressive terms and examining the estimates of the parameters and the residuals. The estimates are not unbiased when the ordinary least squares (OLS) estimation of the ordinary regression is applied to the autoregression. However, the bias is practically trivial when the residuals (errors) are not autocorrelated (Ostrom, 1978; Roberts and others, 1981).

Goodness-of-fit of the model is checked by examining the estimates of autoregressive parameters, R and R^2, and the standard deviation of residuals. For example, if the process is AR(1)

$$x_t = ax_{t-1} + \epsilon_t$$

then the significance of the estimate of the parameter a, say $\hat{a}$, can be tested if ϵ_t is normally distributed, that is, the t-ratio generated by $t = a/\sqrt{\text{Var}[a]}$ (where

Var [a] is the standard error [SE], which can be computed by dividing the standard deviation of the residuals by the square root of the sum of squares of x_{t-1}) is tested against the null hypothesis $a = 0$.

R^2, a common measure of the goodness-of-fit, gives us an idea of the percentage of total variance explained by the autoregression. The standard deviation of the residuals provides similar information when compared with the standard deviation of the original series x_t (Roberts and others, 1981).

The adequacy of the model is determined by the extent to which the underlying assumptions of the model are met. Assumptions for the AR(1), $x_t = ax_{t-1} + \epsilon_t$, are as follows:

1. The relationship between x_t and x_{t-1} is linear
2. $E[\epsilon_t x_{t-1}] = 0$
3. $E[\epsilon_t] = 0$ (mean is zero)
4. $E[\epsilon_t^2] = \sigma_\epsilon^2$ is constant
5. $E[\epsilon_t \epsilon_{t-\tau}] = 0$ (autocovariance is zero)

The behavior of the residuals provides evidence about the extent to which the assumptions are met. For example, assumptions of the model are met when the residuals are uncorrelated with a zero mean and constant variance. In other words, an examination of the ACF should show the autocorrelations of the residuals in the standard error range. The Box and Jenkins (1970) Q-statistic provides a useful measure for this purpose. The Q-statistic for the maximum order M is defined as

$$Q = n \sum_{\tau = 1}^{M} r_\tau^2$$

This is distributed as a chi-square statistic with $M - p$ degrees of freedom (Granger and Newbold, 1977; Box and Jenkins, 1970). The linearity assumption can be examined with a scatter of the residuals against the fitted values.

In goodness-of-fit and adequacy-of-fit are satisfactory, then the impact of the intervention can be estimated. If they are not satisfactory, the above procedure must be repeated with additional terms that may define a more adequate model.

Estimation of the Intervention Effect. Once a parsimonious and adequate model has been identified, an abrupt and permanent intervention effect can be estimated with a step-function dummy variable. The model is expressed as follows:

$$x_t = \sum_{j = 1}^{p} a_j x_{t-j} + bI_t + \epsilon_t$$

where I_t is the intervention variable. $I_t = 0$ when $t < m$ and $I_t = 1$ when $t > m$; m is the point in time when the intervention is introduced. In the second step of this phase, the implementation of the multiple regression gives us the estimate of the intervention parameter b, that is, the estimate is computed by the ordinary least squares method. It should be noted that the regression is implemented with the entire series including the postintervention series. The significance of the b estimate can be tested by the t-ratio generated by $b/\sqrt{\mathrm{Var}[b]}$ against the null hypothesis that $b = 0$.

Goodness- and adequacy-of-fit are again examined visually before the interpretation of the estimated intervention is attempted. This is accomplished by exactly the same operations described above. It should be noted that a statistically significant impact may have little meaning. The statistical significance of the intervention is only one piece of information for the clinical evaluator to consider.

Example

The data analyzed here were taken from Martin and others' (1980) study of the effects of a multiple-component production supervision strategy (PSS) on the productivity of low-functioning retarded adults in a sheltered workshop. The strategy consists of reducing distractions, giving initial instructions, using picture prompts, and having a reinforcement system for productivity.

The hourly production performance rate was measured on a daily basis. On the thirtieth day, the intervention program was introduced to increase the hourly production rate in the morning sessions. There appeared to be some increase in the rate immediately after the introduction of the program. Interestingly, however, the mean hourly production rate in the afternoon sessions seemed to have decreased. In this example, the "contrast" effect on the afternoon production rate will be analyzed. Visual analysis alone was used in the study to assess the intervention impact. The application of the statistical methods described above to these data will demonstrate the utility of these methods for analyzing time series data collected to evaluate the impact of clinical interventions.

Phase I. Visual analysis of Figure 1 reveals the absence of trends in the preintervention and postintervention series. (The statistical software package used for the analysis provided here is IDA, Interactive Data Analysis in Ling and Roberts, 1980). Although there is some up-and-down motion, the variance appears to be constant throughout the series. Further, the series appears to be stationary since the observations tend to vary around a fixed level. Visual inspection, therefore, indicates that a permanent level change can be assumed. The ACF for the preintervention and postintervention series are presented in Figure 2.

Figure 1. Sequence Plot of Production Rate

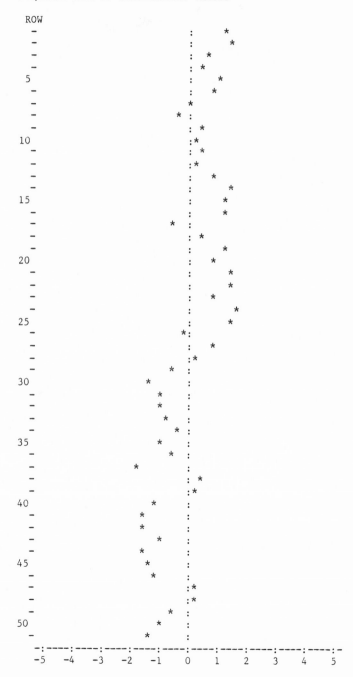

SEQUENCE PLOT OF STANDARDIZED VALUES

Figure 2. Autocorrelation Functions for
Preintervention and Postintervention Series

```
              S.E.
       AUTO- RANDOM
ORDER  CORR. MODEL  -1 -.75 -.50 -.25   0  .25  .50  .75  +1   ADJ.B-P
                    :----:----:----:----:----:----:----:----:
   1   0.315 0.176               +    :      *+              3.183
   2  -0.013 0.173               +    *      +               3.189
   3   0.102 0.170               +    : *    +               3.546
   4  -0.192 0.167             +   *  :      +               4.864
   5  -0.148 0.163             +    * :      +               5.679
   6  -0.165 0.160            +    *  :      +               6.749
   7  -0.333 0.156           *+      :      +               11.28
   8  -0.060 0.153             +    *:      +               11.44
   9   0.174 0.149             +    :   *  +               12.80
  10   0.002 0.145             +    *      +               12.80
                    :----:----:----:----:----:----:----:----:
                    -1 -.75 -.50 -.25   0  .25  .50  .75  +1

              * : AUTOCORRELATIONS
              + : 2 STANDARD ERROR LIMITS (APPROX.)
```

```
              S.E.
       AUTO- RANDOM
ORDER  CORR. MODEL  -1 -.75 -.50 -.25   0  .25  .50  .75  +1   ADJ.B-P
                    :----:----:----:----:----:----:----:----:
   1   0.257 0.199               +    :    * +               1.665
   2  -0.118 0.195               +    * :      +             2.032
   3  -0.325 0.190              +*      :      +             4.976
   4  -0.169 0.185              +   *  :      +              5.811
   5  -0.039 0.179               +    *:      +             5.859
   6  -0.309 0.174              +*      :      +             9.010
   7  -0.153 0.169              +   *  :      +              9.830
   8  -0.047 0.163               +    *:      +             9.913
   9   0.307 0.157               +    :      *              13.75
  10   0.141 0.151               +    :  *  +               14.62
                    :----:----:----:----:----:----:----:----:
                    -1 -.75 -.50 -.25   0  .25  .50  .75  +1

              * : AUTOCORRELATIONS
              + : 2 STANDARD ERROR LIMITS (APPROX.)
```

The preintervention phase has a large first-order autocorrelation that is barely inside the two standard error bands (Figure 2). With the exception only of the seventh order autocorrelation, all the autocorrelations are within the sampling error range. This suggests the series may be stationary. No significantly large autocorrelations can be detected in the postintervention series, suggesting the series may be random. The Box and Jenkins Q-statistic for the series ($Q = 14.62$, $p = .1465$, $df = 10$) supports this observation. To fit the appropriate AR model with the preintervention process, the PACF is examined (see Figure 3). The PACF suggests that the model may be AR(1). However, the high value of the fourth order autocorrelation coefficient could suggest AR(4). First, therefore, a regression with two lagged independent variables (that is, $x_t = 1$, $x_t = 2$, $x_t = 3$, and $x_t = 4$) is examined to make sure that one lagged variable is enough. None of the t-statistics for the second-, third-, and fourth-order autoregressive parameter estimates are significant. Removing the second-, third-, and fourth-order autoregressive terms, we complete the regression with one lagged variable, that is, AR(1) model. The estimate of the first-order autoregressive parameter is significantly different from zero ($t = 1.891$, $p < .10$, $df = 26$). The standard deviation of residuals was reduced to 5.4 from 5.7 in the original series. Thus the model appears to provide a good fit for the data.

The next step is to check the adequacy of the model. First, the ACF of the regression residuals is examined (see Figure 4). None of the autocorrelations fall beyond the two standard error bands. In comparison with the ACF of the original series, the Box and Jenkins Q-statistic also supports the adequacy of the model ($Q = 10.33$, $p = .3244$, $df = 9$). The underlying process of the errors appears to be random. Next, we inspect the scatter plot of the residuals with the fitted values from the regression (see Figure 5). The scatter plot reveals no relation between the residuals and the fitted values, nor does it reveal any serious violation of linearity. Since the diagnostic checks are satisfactory, we proceed with the estimation of the intervention impact.

Phase II. The intervention effect is estimated by a step-function dummy variable. Note that the dependent variable is the whole series (that is, the pre- and postintervention series combined). Examining estimates of the regression parameters, we can see that the estimate of the contrast effect is highly significant ($t = -3.74$, $p < .001$, $df = 47$). The reduction of the hourly production rate from the preintervention phase is approximately ten units. Before concluding that the impact of the intervention is statistically significant, it is necessary to assess the goodness- and adequacy-of-fit.

The estimate of the first-order autoregression parameter is significant ($t = 2.13$, $p < .05$, $df = 47$). The adjusted R^2 is .80 and the standard deviation of the residuals (5.5) is considerably smaller than that of the original series (9.2). The F ratio, an overall measure of the goodness-of-fit of the model, is highly significant ($F = 43.77$, $p < .001$, $df = 2,47$). So, goodness-of-fit appears satisfactory. The ACF of the residuals shows that all autocorrelations fall within the two standard error bands (see Figure 6). The Q-statistic also indicates that the underlying process of the residual series may be random ($Q = 14.35$, $p = .7629$, $df = 19$). The examination of the scatter plot of the residuals and fitted values does not reveal any serious violation of the assumptions of linearity or independence of residuals (see Figure 7). So the adequacy of the model also appears satisfactory from this standpoint.

Figure 3. Partial Autocorrelation Function for Preintervention Series

```
            PARTIAL
ORDER      AUTO COR
                      -1  -.75  -.50  -.25   0   .25  .50  .75   1
                      :----:----:----:----:----:----:----:----:
   1       0.3149     -                    :         *
   2      -0.1243     -                 *  :
   3       0.1638     -                    :    *
   4      -0.3284     -            *       :
   5       0.0859     -                    : *
   6      -0.2775     -             *      :
   7      -0.1389     -               *    :
   8       0.0444     -                    :*
   9       0.1777     -                    :     *
  10      -0.1512     -               *    :
```

Figure 4. Autocorrelation Function of Residuals of Autoregressive Model

```
                   S.E.
          AUTO-  RANDOM
ORDER     CORR.   MODEL   -1  -.75  -.50  -.25   0   .25  .50  .75  +1   ADJ.B-P
                          :----:----:----:----:----:----:----:----:
   1      0.017   0.179              +         *         +            0.9071E-02
   2     -0.155   0.176              +    *    :    *    +            0.7816
   3      0.213   0.173              +         :    *    +            2.310
   4     -0.203   0.169              +    *    :         +            3.745
   5     -0.058   0.165              +        *:         +            3.868
   6     -0.012   0.162             +         *    +                 3.873
   7     -0.294   0.158            *          :    +                 7.332
   8     -0.036   0.154              +        *:    +                 7.385
   9      0.248   0.150              +         :   *+                 10.10
  10     -0.070   0.146              +        *:    +                 10.33
                          :----:----:----:----:----:----:----:----:
                          -1  -.75  -.50  -.25   0   .25  .50  .75  +1
```

```
*  : AUTOCORRELATIONS
+  : 2 STANDARD ERROR LIMITS (APPROX.)
```

Figure 5. Scatter Plot of Residuals with Fitted Values from Autoregressive Model

```
  +                                                    +
  R+                                                   +
  +                                                    +
  E+                                                   +
  +                          1                         +
  S+                      1   1                        +
  +                    1 2    1 2 1 1                  +
  I+           1 1         1 1 1                       +
  -                    1 1        2                    -
  D-                   2  1                            -
  -                 1     1                            -
  U-                   1           1                   -
  -                        1                           -
  A-                                                   -
  -                                                    -
  L-                                                   -
    :---:---:---:---:---:---:---:---:
   -4   -3   -2   -1    0    1    2    3    4
        F    I    T    T    E    D         Y
```

Limitations of the Model

As shown in the example, the analysis of the clinical time series can be conducted relatively easily. However, there are several limitations of the model that must be considered. First, the model may be restricted to the analysis of stationary processes in both pre- and postintervention phases and processes that contain permanent level changes. Secondly, the model may be restricted to series in which the underlying stochastic process can be captured by an autoregressive model. Thus, the given series, as discussed previously, needs to be carefully screened by visually inspecting the process and examining its ACFs and PACFs.

Given these limitations and the availability of various other models, the model presented in this chapter should be considered to be one of the options available for research-practitioners. In practice, time series data often contain trends. When this is the case, the model suggested by Horne and others (1982) may be an appropriate choice. Sometimes the permanent level change cannot be assumed. It may be more appropriate to assume a delayed change or a gradual change. When this is the case, the procedures suggested by Box and Tiao (1975) and McCleary and Hay (1980) may be suitable. Since Marsh and Shibano (1982) identified a substantial portion of clinical time series as sta-

Figure 6. Autocorrelation of Residuals
of Intervention Impact Model

ORDER	AUTO-CORR.	S.E. RANDOM MODEL	-1	-.75	-.50	-.25	0	.25	.50	.75	+1	ADJ.B-P
			:----:----:----:----:----:----:----:----:									
1	0.081	0.137				+	:	*	+			0.3518
2	-0.107	0.136				+	* :		+			0.9719
3	-0.033	0.134				+	*:		+			1.034
4	-0.120	0.133				+	* :		+			1.845
5	-0.055	0.132				+	*:		+			2.017
6	-0.169	0.130				+ *	:		+			3.696
7	-0.179	0.129				+*	:		+			5.641
8	-0.088	0.127				+	* :		+			6.115
9	0.231	0.126				+	:		*			9.503
10	-0.108	0.124				+	* :		+			10.26
11	0.093	0.122				+	:	*	+			10.83
12	-0.059	0.121				+	*:		+			11.07
13	-0.044	0.119				+	*:		+			11.20
14	0.142	0.118				+	:	*	+			12.66
15	-0.065	0.116				+	*:		+			12.97
16	-0.034	0.114				+	.*:		+			13.06
17	-0.071	0.113				+	*:		+			13.46
18	-0.041	0.111				+	*:		+			13.59
19	-0.014	0.109				+	*		+			13.61
20	0.103	0.107				+	:	*	+			14.53

```
        :----:----:----:----:----:----:----:----:
       -1   -.75 -.50 -.25   0    .25  .50  .75  +1
```

```
* : AUTOCORRELATIONS
+ : 2 STANDARD ERROR LIMITS (APPROX.)
```

tionary with a permanent level change, the procedures described here appear to have significant utility. However, differential use of the various models is clearly needed and specific guidelines for selecting an appropriate procedure must be developed.

Conclusions

Recent attention to statistical methods for the analysis of time series data has been limited by the absence of approaches appropriate for the analysis of relatively short, serially correlated data—the types of time series data typically available for evaluating clinical interventions. The approach described in this chapter offers a procedure for the analysis of such data when the pre- and postintervention series are stationary and the intervention has a permanent effect. The procedure is distinguished by the combination of statistical and visual approaches. While fewer observations in shorter series reduce confidence in the model and may increase the possibility of Type II

Figure 7. Scatter Plot of Residuals with Fitted Values
from Intervention Impact Model

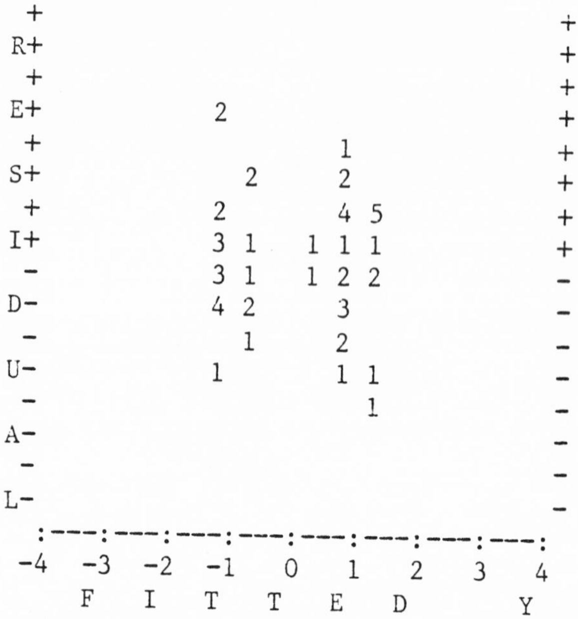

```
    +                                              +
    R+                                             +
    +                                              +
    E+            2                                +
    +                       1                      +
    S+                2     2                      +
    +                2           4 5               +
    I+               3 1    1 1 1                  +
    -                3 1    1 2 2                  -
    D-               4 2      3                    -
    -                  1      2                    -
    U-               1        1 1                  -
    -                           1                  -
    A-                                             -
    -                                              -
    L-                                             -
       :---:---:---:---:---:---:---:---:
      -4  -3  -2  -1   0   1   2   3   4
       F   I   T   T   E   D       Y
```

error, the information provided by statistical analysis in combination with visual analysis gives an additional basis for clinical decision making.

The clinical evaluator is faced with the task of analyzing time series data that may vary in length, presence of trends, and degree of intervention effect. The complexity of the statistical methods required varies with the characteristics of the series. Descriptive studies of clinical time series indicate that they are frequently characterized by stationary series pre- and postintervention with a permanent change induced by the intervention. The analytic approach described provides a straightforward strategy for the analysis of such data.

References

Baer, D. M. "Perhaps It Would Be Better Not to Know Everything." *Journal of Applied Behavior Analysis*, 1977, *10*, 167–172.

Box, G. E. P., and Jenkins, G. M. *Time Series Analysis: Forecasting and Control*. San Francisco: Holden-Day, 1970.

Box, G. E. P., and Tiao, G. C. "Intervention Analysis with Applications to Economic and Environmental Problems." *Journal of the American Statistical Association*, 1975, *70*, 70–79.

Browning, R. M., and Stover, D. O. *Behavior Modification in Child Treatment*. Chicago: Aldine, 1971.

Campbell, D. T., and Stanley, J. C. *Experimental and Quasi-Experimental Designs for Research.* Chicago: Rand McNally, 1966.

Chassan, J. B. *Research Design in Clinical Psychology and Psychiatry.* New York: Appleton-Century-Crofts, 1967.

Gentile, J. R., Roden, A. H., and Klein, R. D. "An Analysis-of-Variance Model for the Intrasubject Replication Design." *Journal of Applied Behavior Analysis,* 1972, *5,* 193–198.

Glass, G. V., Willson, V. L., and Gottman, J. M. *Design and Analysis of Time Series Experiments.* Boulder: Colorado Associated University Press, 1975.

Gottman, J. M. *Time-Series Analysis: A Comprehensive Introduction for Social Scientists.* Cambridge: Harvard University Press, 1981.

Gottman, J. M., and Glass, G. V. "Analysis of Interrupted Time Series Experiments." In T. R. Kratochwill (Ed.), *Single Subject Research: Strateiges for Evaluating Change.* New York: Academic Press, 1978.

Granger, C. W. J., and Newbold, P. *Forecasting Economics Time Series.* New York: Academic Press, 1977.

Hartmann, D. P. "Forcing Square Pegs into Round Holes: Some Comments on an Analysis-of-Variance Model for Intrasubject Replication Design." *Journal of Applied Behavior Analysis,* 1974, *7,* 635–636.

Hartmann, D. P., Gottman, J. M., Jones, R. R., Gardner, W., Kazdin, A. E., and Vaught, R. S. "Interrupted Time-Series Analysis and Its Application to Behavioral Data." *Journal of Applied Behavior Analysis,* 1980, *13,* 543–559.

Hersen, M., and Barlow, D. H. *Single-Case Experimental Designs: Strategies for Studying Behavior Change.* New York: Pergamon Press, 1976.

Horne, G. P., Yang, M. C. K., and Ware, W. B. "Time Series Analysis for Single-Subject Designs." *Psychological Bulletin,* 1982, *91,* 178–189.

Jones, R. R., Vaught, R. S., and Weinrott, M. "Time-Series Analysis in Operant Research." *Journal of Applied Behavior Analysis,* 1977, *10,* 151–166.

Kazdin, A. E. "Statistical Analysis for Single-Case Experimental Designs." In M. Hersen and D. H. Barlow (Eds.), *Single Case Experimental Designs: Strategies for Studying Behavior Change.* Oxford: Pergamon Press, 1976.

Kratochwill, T. R., Alden, K., Dawson, D. D., Panicucci, D. C., Arntson, P., McMurray, N., Hempstead, J., and Levin, J. "Further Consideration in the Application of an Analysis-of-Variance Model for the Intrasubject Replication Design." *Journal of Applied Behavior Analysis,* 1974, *7,* 629–633.

Ling, R. F., and Roberts, H. V. *Users Manual for IDA.* Palo Alto, Calif.: Scientific Press, 1980.

Marsh, J. C., and Shibano, M. "Issues in Clinical Time Series Analysis." Unpublished paper, University of Chicago, 1982.

Martin, G., Pallotta-Cornick, A., Johnstone, G., and Goyos, A. C. "A Supervision Strategy to Improve Work Performance for Lower Functioning Retarded Clients in a Sheltered Workshop." *Journal of Applied Behavior Analysis,* 1980, *13,* 183–190.

McCain, L. J., and McCleary, R. "The Statitstical Analysis of the Simple Interrupted Time-Series Quasi-Experiment." In T. D. Cook and D. T. Campbell (Eds.), *Quasi-Experimentation: Design and Analysis Issues for Field Settings.* Chicago: Rand McNally, 1979.

McCleary, R., and Hay, R. A., Jr. *Applied Time Series Analysis for the Social Sciences.* Beverly Hills: Sage, 1980.

Michael, J. "Statistical Inference for Individual Organism Research: Some Reactions to a Suggestion by Gentile, Roden, and Klein." *Journal of Applied Behavior Analysis,* 1974a, *7,* 627–628.

Michael, J. "Statistical Inference for Individual Organism Research: Mixed Blessing or Curse?" *Journal of Applied Behavior Analysis,* 1974b, *7,* 647–653.

48

Ostrom, C. W. *Time Series Analysis: Regression Techniques.* Beverly Hills: Sage, 1978.

Parsonson, B. S., and Baer, D. M. "The Analysis and Prsentation of Graphic Data." In T. R. Kratochwill (Ed.), *Single Subject Research: Strategies for Evaluating Change.* New York: Academic Press, 1978, pp. 101–166.

Roberts, H., Ling, R., and Bateman, G. "Conversational Statistics." Unpublished manuscripts, Graduate School of Business, University of Chicago, 1981.

Sidman, M. *Tactics of Scientific Research: Evaluating Experimental Data in Psychology.* New York: Basic Books, 1960.

Skinner, B. F. *Science and Human Behavior.* New York: Macmillan, 1953.

Jeanne C. Marsh is assistant professor at the School of Social Service Administration, the University of Chicago.

Matsujiro Shibano is a doctoral student at the University of Chicago.

The use of macro studies and aggregate data can obscure important factors that are critical to an assessment of program effects.

The Effects of Geographic Targeting and Programmatic Change: The Denver Community Development Agency

Alvin H. Mushkatel
L. A. Wilson II

This chapter examines the effect of the decision to geographically target Community Development Block Grant (CDBG) funds in Denver. Confusion in the national community development legislation over who should be the targets of the community development (CD) funds is reviewed in light of the discrepant findings of a number of studies examining the question of "who benefits" from community development resources. Using two different types of analysis, this chapter assesses the impact of the adoption of formal geographic targeting of funds in the Denver CD housing rehabilitation program and offers an explanation for why previous studies addressing the "who benefits" question have produced different findings.

The authors wish to thank Richard McCleary, Lou Weschler, Jeff Martin, Linda Mushkatel, JoAnn Wilson, Colleen Stevens, and Howard Lasus for their assistance.

G. Forehand (Ed.). *New Directions for Program Evaluation: Applications of Time Series Analysis to Evaluation*, no. 16. San Francisco: Jossey-Bass, December 1982.

49

Confusion at the National Level

The Housing and Community Development Act (HCDA) of 1974 and subsequent amendments reflect the ongoing confusion concerning who should and who does benefit from the community development program. The underlying problem contributing to the confusion concerns the criteria used in determining who within a jurisdiction should be recipients of CDBG funds.

Intrajurisdictional targeting of funds "refers to the distribution of activities and benefits within a community" (Dommel and others, 1980, p. 9). Two types of intrajurisdictional targeting exist. The first is social targeting which "seeks to direct many benefits of the block grant to lower income groups" (Nathan and others, 1977, p. 11). This is one of the principal legislative objectives of the HCDA of 1974. The second type of targeting is geographic which "refers to the spatial distribution of funds within a jurisdiction" (Dommel and others, 1980, p. 21). The two types of targeting may lead to a concentration of resources in a few geographic areas. However, it is also possible that the strategies of geographic and income targeting produce a more diverse distribution of benefits. Both a geographical targeting and a social targeting element were incorporated into the final version of the 1974 act.

Consistent with another objective of the HCDA of 1974, namely, encouraging local discretion over the use of the funds, the actual selection of target areas was left up to local governments (Dommel and others, 1980, p. 22). As the CDBG evolved, a growing number of communities concentrated their efforts in transitional neighborhoods (U.S. Department of Housing and Urban Development, 1978, p. 44; Dommel and others, 1978, p. 243; Nathan and others, 1977, p. 33; Rosenfeld, 1979, pp. 448–457). During the Ford administration, HUD seemed willing to allow this concentration of funding in transitional areas to continue, using the dual requirements of the HCDA as a justification.

With the arrival of the Carter administration, HUD became concerned that geographic targeting combined with local discretion had resulted in too many communities "spreading" their resources to families with greater than low or moderate incomes. Shortly after taking office, HUD Secretary Harris testified before a House subcommittee and placed local communities on notice that HUD was about to shift its emphasis toward social targeting. Two events made this shift apparent. First, HUD proposed that CDBG funds principally benefit "low and moderate income families" (Housing and Community Development Act of 1977, P.L. 95–120, Section 104 B)(1); Dommel and others, 1980, p. 16). This wording marked a change from the original 1974 legislation, which designated the beneficiaries to be of low *or* moderate income (Keating and LeGates, 1978, p. 709). Second, Harris announced that HUD would initiate a "substantive review of community utilization of CDBG funds"

(Nathan and others, 1977, p. 15). It was HUD's intent to adopt a 75-25 rule in evaluating local programs. According to this guideline, at least 75 percent of the locale's CDBG funds would be earmarked for "low and moderate income families" (Dommel and others, 1980, p. 17; Van Horn, 1979, pp. 129–130). Despite these actions, Congress created confusion over targeting with the amendments passed in 1978 requiring HUD to retreat in applying the 75-25 rule.

Because of HUD's apparent indecisiveness and the weakening by Congress of HUD's position on social targeting, confusion existed concerning how to determine the principal beneficiaries of CDBG funds. Both social and geographic targeting were accepted, so individual communities were free to choose either low- and moderate-income families or low- and moderate-income areas as targets.

Evaluation of CDBG — Who Benefits?

As a result of the duality in the CD act itself, one of the major problems encountered by evaluative studies assessing "who benefits" from CDBG funds is identifying what unit of analysis to utilize. The focus will vary depending on whether a community has adopted a social or a geographical targeting approach to fund allocation. In addressing the question of "who benefits?" some studies have focused on geographical targets as the unit of analysis, while others have used social targets as the unit of analysis.

Compounding the unit-of-analysis issue is the problem of operationally defining what constitutes low and moderate income (Keating and LeGates, 1978, p. 712). Some studies (for example, the Brookings Monitoring Series; see Dommel and others, 1978, 1980; Nathan and others, 1977) employ HUD's definition. Low-income families are defined as those earning less than 50 percent of the median income of their SMSA, and moderate-income families are those earning between 50 percent and 80 percent of the median income (Nathan and others, 1977, pp. 306–307). Using the SMSA medians tends to inflate the proportion of funds going to low- and moderate-income neighborhoods within cities. The Standard Metropolitan Statistical Area (SMSA) figures are higher than the central-city medians, so more tracts and families from central cities than from outlying areas fall into low- and moderate-income categories (Rosenfeld, 1977, p. 32). A further frustration in doing evaluative studies is that some communities differentiated low- from moderate-income tracts and families, while others did not. For example, HUD's analysis did not distinguish between low- and moderate-income areas (Keating and LeGates, 1978, p. 711).

After making a thorough analysis of the differences in methodologies and in the way benefits are allocated to geographical targets, Van Horn (1979,

p. 128) concludes that there has been a slight decline for low- and moderate-income areas as a result of the shift away from categorical programs to the CDBGs. The share of benefits from CDBG funds going to "lower" income areas varied from one-half to three-fifths of the total funds. A general conclusion of this and other studies is that the proportion of funds allocated to low- and moderate-income areas declined for each year the block grant program continued (Keating and LeGates, 1978, p. 710).

However, looking at social targets, Dommel and others (1980, p. 161) have suggested that the proportion of benefits allocated to low- and moderate-income families has increased from 54 percent in the first year of the block grant program to 62 percent in the fourth year. Whereas studies looking at geographic targets found a decline in benefits to low- and moderate-income spatial units, the Brookings study, focusing on social targets, finds an increase in the proportion of resources allocated to low- and moderate-income families.

The discrepancies among these studies indicate a need for more thorough analysis of a single program to determine the effects of geographic targeting, rather than reliance on large macro studies that combine all CD activities and programs. Only by disaggregating the data can one observe the different targeting effects and the dynamics of the process. This study analyzes one program, a housing rehabilitation program in Denver, to determine what, if any, was the effect of formally adopting geographic targeting on the allocation of CDBG funds for rehabilitation. As we will show, the use of macro studies and aggregate data can obscure important factors that are critical to an assessment of the program effects.

Denver's Community Development Rehabilitation Program (CDRP)

Prior to 1978 the housing rehabilitation program operated by Denver's Community Development Agency (CDA) was a grant program. The precise criteria for awarding the grants were never made explicit, and it seems probable that political and standard social targeting considerations influenced the rehabilitation allocations.

In late 1977, a task force was established to revise the CDRP and to establish clear criteria for awarding rehabilitation grants and loans. In addition, the city planned to "target rehabilitation resources into certain designated areas of the city" ("Community Development . . .," 1978, p. 1). The CDA was going to adopt geographic targeting as a formal guideline for its allocations for the rehabilitation program.

The resulting new program contained both social and geographic targeting components. Although families would have to be of low or moderate income to obtain rehabilitation monies, where they lived would also influence the likelihood of their receiving funds. It is this combination of targeting that we alluded to earlier and that might explain, in part, why low- and moderate-income families are capturing a greater percentage of the funds while at the same time low- and moderate-income areas are receiving less support.

The decision to establish criteria for geographically targeting areas within the city was aided by events in Washington in 1977. In August 1979, CDA planners in Denver expressed their beliefs that HUD was about to crack down on communities that had not established clear criteria for distributing funds or those that could not meet the proposed 75–25 rule. Therefore, Denver initiated a process that would result in the development of formal criteria for selection of target areas.

Based on characteristics of the census tracts within them, five target areas were selected. According to the CDA's final rankings of census tracts, two of the five target areas come from the highest-need areas. The remaining three target areas are from areas of less need ("Community Development...," 1978, Appendix B). At least three of these areas, then, appear to fall into what have been described loosely as transitional areas rather than areas of greatest need. One of these three transitional areas also exhibits an income level above low- and moderate-income areas as defined by HUD (80 percent of the median income for the SMSA). Only one of the areas in the target group had a median income level fitting HUD's description of low income (50 percent of the SMSA median income). In short, the development of criteria to select target areas resulted in choosing one middle-income area and two transitional areas along with two areas from the greatest-need category.

Given the nature of the target areas selected, questions arise concerning whether or not the CDA was able to channel the money to these target areas and, if so, where this money came from.

The Data and Methods of Analysis

The objective of this study is to determine whether or not the adoption of formal geographic targeting in Denver has led to a shift of funds away from low- and moderate-income tracts to middle- and upper-income tracts. The analysis focuses on geographic, sociospatial units, that is, census tracts. The data do not include the income levels of all individuals receiving loans/grants frm the CDA; however it is assumed that after CDRP implemented income eligibility requirements, low- and moderate-income individuals received the majority of funding.

The data consist of records for every loan/grant made for the CDRP from fiscal year 1974–1975 through June of 1980. Data were provided by the Denver Urban Renewal Authority, which administers this CDA program. The records consist of the dollar value of each loan, location of the housing unit receiving the loan, completion date of each project, and the contractor used to rehabilitate the home.

We have located the sites receiving each of these loans into their respective census tracts. Census tracts are identified as low or moderate income if they fall below 80 percent of the median family income for the SMSA (HUD's definition). Basing our definition on a cutoff level for the entire SMSA is generous to the CDA, because the city's median family income is lower than that for the SMSA (SMSA, $10,777; city, $9,654).

In over five years of the program, the CDA claims it has spent over $18 million. Yet, when the administrative costs are filtered out and some apparent loss of records is taken into account, the total spent appears to be just over $8.25 million for rehabilitation loans/grants. Because our objective is to determine the effect of formal geographic targeting on the pattern of funds distributed to census tracts, the data are divided into two time periods. The time periods are divided by the CDA's formal adoption of criteria to target funds and the issuance of their task force report in May of 1978. The period prior to the program change runs from 1974–1975 through April 1978. Records indicate that during this time period the CDA allocated $2,967,980 for rehabilitation loans and grants. The period following the implementation of targeting (posttargeting period) runs from May 1978 through August 1980 (last available data). During this posttargeting period the CDA allocated $5,254,903 for rehabilitation.

The findings reported in Table 1 allow us to examine the effects of geographic targeting in Denver. The targeting of funds after May 1978 appears to have increased the proportion of funds to target areas. Target areas increase their proportion from 12.44 percent in the pretargeting period to 14.97 percent after the formal decision to target funds was made. Even more interesting is the fact that areas above 80 percent of the SMSA income increase their proportion of funding by almost 9 percent from 27.98 percent to 36.30 percent.

Where did the funds to increase the proportional allocations of the target area and areas with above 80 percent of the median income come from? Table 1 demonstrates that the poorer areas of the city (below 80 percent of the median-income level) lost a substantial proportion of the percent of funds received as a result of the formal adoption of geographic targeting. Almost 11 percent of the proportion of money allocated to these needy areas in the pretargeting period is lost in the new program. In short, the poorer areas of the city are losing a substantial proportion of their allocation to pay for proportional increases among the target and better off areas of the city.

The above analysis—based as it is upon a simple dichotomy between pretargeting and posttargeting periods—may well obscure the nature and magnitude of change in loan granting behavior as a result of change in official policy. As Cook and Campbell (1976) have pointed out, one is well advised to

Table 1. Distribution of CDRP Funds Before and After
May 1978 Program Changes by Income of Census Tracts
with Target Areas Reported Separately

	Period before May 1978 Program Changes	Period after May 1978 Program Changes
Tracts above 80% of SMSA Median Income N = 52	$ 830,692 (27.98%)	$1,908,023 (36.30%)
Tracts below 80% of SMSA Median Income N = 31	1,768,033 (59.57%)	2,559,959 (48.71%)
Target Area N = 6	369,255 (12.44%)	786,921 (14.97%)
TOTAL	$2,967,980	$5,254,903

disaggregate pretargeting and posttargeting data and carefully examine the effects of the intervention or policy change on one's observations. The data were disaggregated for the three groupings of census tracts by first, the number of loans granted to each of the areas, and second, the average size of these loans in the natural log metric.

A cursory inspection of these data reveals that some change appears to have occurred in almost all of the referenced series. The important questions, of courses, are (1) when did the change take place? (2) what is the nature of that change? and (3) is the observed change statistically significant (other than what would have been expected by chance)?

The disaggregated data have been analyzed using the stochastic process autoregressive integrated moving average (ARIMA) models as developed by Box and Jenkins (1976). The specific software package used for this analysis is the one developed by David Pack (1977). The procedures used in conducting the analysis of these data are those that have been suggested by McCain and McCleary (1979, p. 267). That is, the three-step procedure of identification, estimation, and daignosis has been adopted for this research. Consistent with the recommendations of McCain and McCleary (1979) and McCleary and Hay (1980), the comparative adequacy of competitive models is based upon the twin concerns of statistical significance and parsimony.

An examination of the autocorrelation functions (ACF) and partial autocorrelation functions (PACF) of the identification program provided by Pack (1977) reveals that, for each of the time series analyzed here, an ARIMA (1,1,0) noise model appears to be appropriate. Therefore, the specific noise model to be estimated in the second stage of the analysis is:

$$Y_t = \frac{a_t}{(1 - \theta B)(1 - B)}$$

The adequacy of this noise model is evaluated in three different ways. First, the Q-statistic taken from the estimation program is evaluated for statistical significance in which the null hypothesis is that the data constitute a white noise process. Second, the specific autoregressive parameter estimated for the process is evaluated for statistical significance. In this instance, the null hypothesis is that the autoregressive component is not significant. The third test of adequacy is found in the evaluation of the significance of individual autoregression functions taken for each of the specified lags. When these values are twice their standard error, they are assumed to be significant and deserving of further attention (that is, one should give thought to reidentifying the original noise model).

The results of this process of identification and estimation are presented in Tables 2, 3, and 4. In each of these tables, results of the analysis for both May 1978 and April 1977 intervention points are presented. Focusing initially on the results presented in each of the tables for May 1978, it is clear that the ARIMA (1,1,0) model is appropriate for each of the time series. The Q-statistic for both average loan size and number of loans data series is statistically insignificant (using the .05 confidence level) for each of the areas. Consequently, one can assume the processes now approximate white noise.

The autoregressive component of the ARIMA (1,1,0) model is statistically significant in each of the areas for both data series as well. In each instance, the calculated value of the statistic is greater than 1.96 (.05 confidence level). The overall evaluation must be, therefore, that the ARIMA (1,1,0) model is the appropriate one for each of these series.

Evaluation of the intervention component for each of the data series begins with an examination of the first-order transfer function. This first-order transfer function is:

$$f(I_t) = \frac{\omega_0}{1 - \delta B} I_t$$

In evaluating the appropriateness of the first-order transfer function, one needs to be concerned with the statistical significance of the individual parameters (both delta and omega) as well as with the fact that the delta parameter fits within the bounds of system stability (McCleary and Hay, 1980, p. 155).

Table 2. Target Tracts

May 1978 Intervention	Estimate	Lower	Upper	t
Average Loan Size				
Q-Statistic	25.85			
AutoRegressive	-.439	-.653	-.225	4.027
Delta	.420	-9.422	10.264	.83
Omega	.045	-.593	.684	.138
Number of Loans				
Q-Statistic	27.62			
AutoRegressive	-.420	-.640	-.200	3.74
Delta	-.343	-.9818	9.130	.075
Omega	-.131	-1.416	1.154	.199

April 1977 Intervention	Estimate	Lower	Upper	t
Average Loan Size				
Q-Statistic	20.86			
AutoRegressive	-.450	-.662	-.237	4.147
Delta	-.454	-.989	.081	1.669
Omega	1.126˙	.445	1.808	3.244
Number of Loans				
Q-Statistic	29.73			
AutoRegressive	-.360	-.591	-.129	3.076
Delta	-.543	-2.627	1.540	.511
Omega	-2.507	-9.152	4.136	.739

NOTE: Evaluating the Q-statistic with 29 degrees of freedom, a Chi-Square value of 42.56 is signficant at the .05 level.

Looking first at the statistical significance of the various omega parameters reported for the May 1978 intervention point in each of the three areas, it is apparent that a statistically significant shift in level of each series is found only for the number of loans in the tracts under 80 percent of the median income. That is, using this intervention date, no statistically significant shift in level of either average loan size or number of loans is found in either the target tracts or those tracts above 80 percent of the median income.

Turning attention to the results reported for an April 1977 intervention date, statistically significant omega parameters are found for average loan size in both target tracts and tracts above 80 percent of the median income. Interestingly, no statistically significant shift parameters are reported for tracts

Table 3. Tracts Above 80 Percent of Median Income

May 1978 Intervention	Estimate	Lower	Upper	t
Average Loan Size				
Q-Statistic	15.75			
AutoRegressive	-.415	-.634	-.196	3.738
Delta	-.637	-.016	.396	1.208
Omega	-.372	-1.003	.258	1.158
Number of Loans				
Q-Statistic	29.01			
AutoRegressive	-.445	-.673	-.217	3.836
Delta	-.493	-2.557	1.571	.468
Omega	-4.681	-16.117	6,753	.802

April 1977 Intervention	Estimate	Lower	Upper	t
Average Loan Size				
Q-Statistic	21.92			
AutoRegressive	-.473	-.687	-.259	4.339
Delta	-.448	-1.086	.190	1.378
Omega	.859	.249	1.468	2.764
Number of Loans				
Q-Statistic	31.57			
AutoRegressive	-.450	-.664	-.234	4.128
Delta	.671	.089	1.261	2.229
Omega	1.526	-2.226	12.758	.266

NOTE: Evaluating the Q-statistic with 29 degrees of freedom, a Chi-Square value of 42.56 is signficant at the .05 level.

under 80 percent of the median income when using the April 1977 intervention date.

In evaluating the significance of the intervention points, initial use was made of first-order transfer function. The delta parameter in this transfer function is a rate of change parameter. Had a large and statistically significant delta parameter been reported for a specific intervention, the effect of the intervention would occur quickly (reach its asymptotic level in a few observations). In contrast, had a small and statistically significant delta parameter been reported, the effect of the intervention would have been gradual (taking

Table 4. Tracts Under 80 Percent of Median Income

May 1978 Intervention	Estimate	Lower	Upper	t
Average Loan Size				
Q-Statistic	25.93			
AutoRegressive	-.299	-.525	-.072	2.600
Delta	.777	-.012	2.780	.761
Omega	-.055	-.433	.322	.286
Number of Loans				
Q-Statistic	29.46			
AutoRegressive	-.248	-4.484	-.011	1.878
Delta	.313	-.390	1.017	.871
Omega	-16.883	-31.484	-2.282	2.266

April 1977 Intervention	Estimate	Lower	Upper	t
Average Loan Size				
Q-Statistic	31.24			
AutoRegressive	-.337	-.566	-.108	2.905
Delta	.575	-.465	1.615	1.083
Omega	.246	-.233	.725	1.008
Number of Loans				
Q-Statistic	41.04			
AutoRegressive	-.280	-.508	-.052	2.413
Delta	.730	.073	1.386	2.185
Omega	.501	-.425	1.429	1.059

NOTE: Evaluating the Q-statistic with 29 degrees of freedom, a Chi-Square value of 42.56 is signficant at the .05 level.

many observations to reach its asymptotic level). As can be seen from the results presented in Tables 2 through 4, none of the delta parameters reach statistical significance when the omega parameter is statistically significant.

The diagnosis phase of this investigation, therefore, leads one to reformulate the transfer function as one that is of a zero order. The zero-order transfer function is:

$$f(I_t) = \omega_0 I_t$$

Results of the analysis using the zero-order transfer function are presented in Table 5. As can be seen through comparison of the values reported in Table 5

Table 5. Target Tracts: Zero Order Transfer Function

April 1977 Intervention	Estimate	Lower	Upper	t
Average Loan Size				
Q-Statistic	24.80			
AutoRegressive	-.471	-.678	-.264	4.485
Omega	.950	.333	1.568	3.015

TRACTS ABOVE 80% OF MEDIAN INCOME
Zero Order Transfer Function

April 1977 Intervention	Estimate	Lower	Upper	t
Average Loan Size				
Q-Statistic	22.28			
AutoRegressive	-.464	-.675	-.253	4.336
Omega	.701	.150	1.252	2.494

TRACTS UNDER 80% OF MEDIAN INCOME
Zero Order Transfer Function

May 1978 Intervention	Estimate	Lower	Upper	t
Number of Loans				
Q-Statistic	30.97			
AutoRegressive	-.257	-.490	-.025	2.177
Omega	-17.385	-31.995	-27.745	3.289

with those reported in Tables 2 through 4, this more parsimonious model has resulted in some modest changes in the reported values of the omega shift parameters. (McCleary and Hay [1980] note that extreme outliers may bias the findings of one's analysis. To control for the possible effect of the major outlier occurring in the target tracts for the average loan size data, reanalysis with a "Windsorized" estimate of the value was substituted. No change was noted.)

What, then, might be said about the impact of these various changes on the average loan size and number of loans noted in the various areas? The most straightforward interpretation of the information presented in Table 5 can be given to that presented for tracts under 80 percent of the median income. The data analyzed in this table are in the raw metric, which means, in turn, that the shift parameter is reported in the raw metric. The apparent effect of the policy change is an immediate drop in the level of the series by between seventeen and eighteen loans a month for this area. A clear negative effect, therefore, is noted for this area in terms of the May 1978 intervention.

The shift parameters for the average loan size in both target tracts and tracts above 80 percent of the median income are a little less straightforward. In these instances, the term $e^{(0)}$ is interpreted as the "ratio of the postintervention series level to the preintervention series level" (McCleary and Hay, 1980, p. 174). Hence, for the target tracts one can say that the average loan size (using the April 1977 intervention date) increased by 61.3 percent in the postintervention time period. The average loan size in the tracts above 80 percent of the median income increased by 50.4 percent in the postintervention time period.

It should be emphasized that the nature (but not direction) of the three significant impacts noted is the same: there is an immediate shift in the level of the time series. This step-function effect — which is denoted by the lack of statistical significance of the delta parameters — means that the policy change was abrupt, not gradual, in its implementation.

The most striking difference between the time series and aggregate analyses is identification of the *time* of policy change or innovation. The time series data clearly indicate that the policy change regarding target and over 80 percent of the median-income areas predated the formal notice of decision (May 1978). In fact, using this date with the time series data, one finds no statistically significant change in loan granting behavior (as measured by number of loans and average size of loans) in the under 80 percent of median-income areas, and the impact of this change on the reduction of the number of loans to this area.

That there is inconsistency in assessment of change for the aggregate and time series should cause the time series analyst no alarm. It is easy to mask the timing of change through the aggregation of time series data (see, for instance, Cook and Campbell, 1976, p. 276). Obviously, such insensitivity to the issue of time of intervention should give the aggregate data analyst some concern.

The use of time series analysis has allowed us to understand how the effects shown by the aggregate analysis were generated. The proportional increases in funds to the target tracts above 80 percent of the median income were accomplished by increasing the amount of the average loan to these areas. The proportional loss of funds in tracts below 80 percent of the median income was accomplished by reduction in the number of loans to these areas.

The time series analysis made it possible to identify the dynamics by which the observed effects were achieved.

The implications of this study are important for the larger aggregate studies of the use of CDBG funds as well as for policy evaluation studies in general. With respect to larger studies of the CDBG, this study suggests that such large scale efforts as evaluation may mask the effects of a program and the dynamics involved. The study by Brookings, which does not distinguish changes in the ways programs are implemented, fails to determine the effect of a specific program and in aggregating across cities and program areas is likely to be misleading in its answer to the question of "who benefits" from CDBG funds.

Our study indicates the importance of what is selected as the intervention date in evaluative studies. It may be misleading to utilize a program intervention date provided by the involved agency; it may alter the findings of the evaluative effort. As has been demonstrated in the case of the CDA in Denver, if one uses the DCA date of May 1978, the dynamics identified as producing the change in allocative proportions to the three types of tracts are very different than if one uses the April 1977 date.

A possible explanation for the discrepancy between the agency identified intervention date and the actual intervention date is that it may be that an agency decides to change its allocation scheme and in so doing a series of steps are needed until it becomes fully operational. Hence, one might observe an initial dramatic change in the allocations followed by less dramatic changes until the policy is fully operational and adopted by the appropriate agency of government.

The important point is that no matter what is the reason for discrepancy, efforts that do not employ time series data may mislead the investigation and probably underestimate the effect of the impact of a program change. Using aggregate studies and accepting the agency's date of intervention may result in underestimating the program effect and the dynamics associated with the impact. However, studies that disaggregate the data and attempt to determine other probable dates of intervention may open themselves to the charge of blatant empiricism. Our approach has been to use the agency's date of intervention and to consider the policy-relevant events at the local and national level for an obvious point in time at which the program may have been altered. In the case of Denver, this has allowed us to determine both the effect of geographic targeting and the process responsible for these results.

References

Box, G. E. P., and Jenkins, G. M. *Time Series Analysis: Forecasting and Control.* (Rev. ed.) San Francisco: Holden-Day, 1976.

"Community Development Rehabilitation Program 1978" Unpublished report, Community Development Agency, City and County of Denver, 1978.

Cook, T. D., and Campbell, D. T. "The Design and Conduct of Quasi-Experiments and True Experiements in Field Settings." In M. Dunnette (Ed.), *Handbook of Industrial and Organizational Psychology.* Chicago: Rand McNally, 1976.

Dommel, P., Back, V., Liebschutz, S., Rubinowitz, L., and others. *Targeting Community Development.* Washington, D.C.: U.S. Government Printing Office, 1980.

Dommel, P., Nathan, R., Liebschutz, S., Wrightson, M., and others. *Decentralizing Community Development.* Washington, D.C.: U.S. Government Printing Office, 1978.

Housing and Community Development Act of 1974. Public Law 93-383.

Housing and Community Development Act of 1977. Hearings Before the Subcommittee on Housing and Community Development of the Committee on Banking, Finance and Urban Affairs, Part 1. Washington, D.C.: U.S. Government Printing Office, 1977.

Keating, D., and LeGates, R. "Who Should Benefit from the Community Development Block Grant Program?" *Urban Lawyer,* 1978, *10* (4), 701-736.

McCain, L. J., and McCleary, R. D. "The Statistical Analysis of the Simple Interrupted Time-Series Quasi-Experiment." In T. D. Cook and D. T. Campbell (Eds.), *Quasi-Experimentation: Design and Analysis Issues for Field Settings.* Chicago: Rand McNally, 1979.

McCleary, R. D., and Hay, R. A., Jr. *Applied Time Series Analysis for the Social Sciences.* Beverly Hills: Sage, 1980.

Nathan, R., Dommel, P., Liebschutz, S., Morris, M. D., and Associates. *Block Grants for Community Development.* Washington, D.C.: U.S. Government Printing Office, 1977.

Pack, D. J. "A Computer Program for the Analysis of Time Series Models Using the Box-Jenkins Philosophy." Hatboro, Pa.: Automatic Forecasting Systems, 1977.

Rosenfeld, R. "National and Local Performance in Community Development Block Grants: Who Benefits?" Paper presented at the annual meeting of the American Political Science Association, Washington, D.C., 1977.

Rosenfeld, R. "Local Implementation Decisions for Community Development Block Grants." *Public Administration Review,* 1979, *39,* 448-457.

U.S. Department of Housing and Urban Development. *Community Development Block Grants: Third Annual Report.* Washington, D.C.: U.S. Government Printing Office, 1978.

Van Horn, C. *Policy Implementation in the Federal System.* Lexington, Mass.: Lexington Books, 1979.

Alvin H. Mushkatel and L. A. Wilson II are associate professors at the Center for Public Affairs, Arizona State University.

A time series analysis is used to evaluate a change in
management style at a federal correctional institution.

Functional Unit Management: Organizational Effectiveness in the Federal Prison System

Michael Janus

As early as 1975, in response to the growing skepticism about the effectiveness of the rehabilitative model, the director of the federal prison system, Norman Carlson, had defined the goal of his organization as humane control, with ample opportunity for the inmate to participate in self-rehabilitative therapy if he or she so desired (see Lejins, 1975, p. 59). The concept of human control connotes a situation in which inmates are treated fairly and justly. Nevertheless, the institutional environment requires a degree of control sufficient to protect both inmates and staff. Additionally, Carlson (1976) set forth the policy that inmates be given every opportunity to participate in treatment-oriented programs on a voluntary basis.

The adoption of the goal of humane control and the concurrent rejection of the goal of coerced therapy presents the federal prison system with some potentially troublesome organizational problems. These problems include the delineation of the role of both the inmate and the professional

The views presented in this article represent those of the author, and not necessarily those of the Department of Justice, or the federal prison system.

G. Forehand (Ed.). *New Directions for Program Evaluation: Applications of Time Series Analysis to Evaluation*, no. 16. San Francisco: Jossey-Bass, December 1982.

correctional staff in the institution, and the maintenance of adequate institutional control by correctional authorities. Partially in response to these problems, the federal prison system has recently adopted an organizational style designated functional unit management.

Functional unit management (FUM) can best be described as a management technique that decentralizes the organizational structure of a prison. In effect, the implementation of this style restructures a large institution into several mini-institutions (units), which coexist in the same prison. "A unit is a small, self-contained inmate living and staff office area which operates semiautonomously within the confines of the larger institution" (Lansing and others, 1977, p. 43). Clerical, educational, and psychological departments are no longer independently responsible for their respective functions, but serve as resource centers for their representatives within each unit. The unit represents a relatively autonomous suborganization composed ideally of 50–250 inmates and their respective staff. A change to unit management is represented by: (a) a change in the organizational structure so that the unit manager assumes many of the authoritative responsibilities previously performed by department heads; (b) physical relocation of staff to unit offices; and (c) a change in programming activity to unit emphasis rather than institutional emphasis.

In contrast to the specialized function required in a traditional bureaucratic structure (for example, the inmate traditionally goes to the department of psychology for counseling), the unit staff must now take on a wider role by providing more direct services to the inmate in the unit. By thus involving the staff in the day-to-day lives of inmates, it is hoped that control without coercion may be achieved.

The stated goals of unit management reflect the ultimate goals of the federal prison system (1) to establish a safe, humane environment that minimizes the detrimental effects of confinement and (2) to provide a variety of counseling, social, educational, and vocational training opportunities and programs that are most likely to aid inmates in their successful return to the community.

The contributions that unit management is expected to make toward the attainment of these goals rest primarily on the assumed advantages of staff-inmate and inmate-inmate familiarity. The major advantage of unit management is that:

It increases the frequency of contacts and the intensity of the relationship between staff and inmates, resulting in:
a. better communication and understanding between individuals
b. more individualized classification and program planning
c. more valuable program reviews and program adjustments of problems before they reach critical proportions

d. development of common goals which encourage positive unit cohesiveness

e. generally, a more positive living and work atmosphere for staff and inmates, and

f. more efficient accountability and control of inmates [*Unit Management Manual,* 1980, p. 2].

With the current de-emphasis on rehabilitation as a primary goal and the increasing stress on control, this research focuses on the goal of humane control. More specifically, I attempt to measure the effects of the implementation of functional unit management on the goal of humane control within the institutional setting. This work does not attempt to draw conclusions about the effect that humane control ultimately has on the crime problem in society. A relevant perspective on the relationship between just, humane control in the prison setting and its effects on the reduction of crime is developed by Fogel (1975).

Previous Evaluations of Functional Unit Management

In 1975, Bogan, Karacki, and Lansing surveyed eleven previous attempts by the federal prison system to evaluate the effects of unit management. The dependent variables used in these studies included personality profiles, recidivism rates, and incident data. Although there were some inconsistencies, in general the results of these evaluations indicated positive results for functional unit management. Indeed, the positive results from these early evaluations were a major factor in the decision to introduce unit management as the organizational style of the federal system (*Unit Management Manual,* 1980, p. 5). Bogan and others (1975) summarized: "The research results to date indicate that functional unit management has generally positive effects on institutional variables; its effects on personal adjustment or postinstitutional variables have yet to be demonstrated. It should be remembered that functional unit management as a program and management tool is, in most institutions, in its infancy. A longer period of time and a complete research effort is required to assess the true impact of functional unit management" (p. 17).

Previous evaluations of unit management have been flawed by both methodological and conceptual problems. Most investigations have used a single pre- and a single post-FUM implementation measurement. This is seen in the studies involving the Correctional Institution Environment Scale (CIES) (Lee, 1980). The CIES is an evaluative instrument for correctional institutions that utilizes attitudinal measures of satisfaction with environment (see Moos, 1975). In some studies, measures of incident rates and inmate or staff attitude have been aggregated and averaged in a single pre- and post-

FUM time frame (Rowe and others, 1976). This approach threatens the validity of the research by failing to investigate the dynamics of the effects of unit management. For instance, unit management may cause a temporary increase in positive attitudes, followed by a resumption of earlier levels of institutional climate (a Hawthorne effect). Therefore, the reported effects of the implementation of FUM would vary depending on the time frame in which the survey was administered.

Additionally, an attempt to measure the effects of a complex organizational arrangement such as unit management, while ignoring concurrent organizational developments such as personnel or inmate population changes, may be misleading. Of all of the studies reported to date, none has attempted to control for such developments.

Finally, some of the more comprehensive evaluations were completed earlier in the development of unit management. Consequently, those evaluations were concerned primarily with the effects of unit management as a classification or rehabilitation technique within specialized units, rather than as a total reorganization of an institution following the more recently adopted policy of humane control.

The methodology described in the following section was developed with the weaknesses of previous research efforts in mind. The research setting was one in which behavioral data is available over time. The aggregate time periods (weeks) are small enough to observe the dynamics of fluctuations in the dependent variables in response to the introduction of functional unit management. Also, this type of data aggregation made it possible to separate the effects of unit management from other events occurring in the institutional setting. Other events such as major personnel or policy changes were documented and separated from the effects of unit management to the extent possible.

The time period for the study (1975–1977) was one in which "humane control" had been established as a primary goal in the federal prison system. Therefore the variables used to measure organizational effectiveness were selected to reflect this goal rather than the goals of classification or rehabilitation.

The research setting for this project was the Federal Correctional Institution (FCI) at Lompoc, California. FCI Lompoc was chosen because it is a major institution (approximately one thousand inmates), and Lompoc shifted to unit management during the era of humane control as a primary goal for the federal prison system (1976). Except for a drug unit, inmate assignment to units was on an availability basis.

At the time that Lompoc underwent the transition to unit management, it was a medium-security-level institution. This security designation at the time of the implementation of functional unit management characterizes Lompoc as an average federal prison, on a par with most state institutions. This similarity enhances the generalizability and external validity of the study.

Optimally, in a pure experimental design, a second institution, or a portion of Lompoc that was not undergoing a change to unit management, would be used as a control. However, no similar data are available from other institutions in the same historical time frame, and the entire institution changed to unit management at about the same time. Although some units were physically remodeled before others, it would be fallacious to use a section of Lompoc as a control. The lack of a comparison or control group precludes the use of a pure experimental design.

The Variables

Incidents. An incident is recorded if a correctional officer believes that an action by an inmate is sufficiently severe to warrant official action. The frequency of incidents, taking into account their severity, which is also recorded, will be used as a general indicator of both the tenseness of the environment at the institution and the need for the use of official channels to maintain control. It is assumed that if unit management is achieving the goal of humane control, both the number and seriousness of reported incidents should decrease.

It was anticipated that the reporting of less serious incidents might temporarily increase during the early stages of FUM implementation due to increased intensity of supervision. However, more serious incidents were expected to decline in number. For this reason, incidents were dichotomized according to seriousness, and analyzed separately.

Administrative Remedies. The administrative remedy is the inmates' channel for registering a complaint of unfairness against any correctional worker or any decision by a correctional worker. It was assumed that as the quality and quantity of inmates' interaction with staff, especially decision-making staff, increased, the number of times that inmates must rely on official channels of grievance would decrease. Therefore, the implementation of unit management should cause a decrease in the number of administrative remedies filed.

Both of the above variables are unobtrusive measures (Webb and others, 1966, pp. 53–87). That is, they were not originally collected for experimental purposes. Aside from the relatively low cost of acquisition, Webb and others (1966, p. 53) point out that a common benefit to be derived from this type of data is its "nonreactivity." It is highly unlikely that either of the above variables were being altered in anticipation of a study of the effectiveness of unit management.

Also, since the use of both incident reports and administrative remedies are discretionary activities, it is possible that a change in management style will alter the nature of the reporting rather than the quality of humane control. Cook and Campbell (1979, p. 59) refer to this threat of confounding as a problem of "construct validity." This threat represents a problem not unlike problems associated with crime-reporting rates. That is, the primary method of measuring the frequency of incidents rests with their official recognition. In this study it was assumed that both inmates and staff frequently resolve these problems (inmate misconduct or complaints against the system) informally, and the attainment of this informal resolution is, in and of itself, a reflection of more effective and humane control.

In his analysis of crime reporting behavior, Black (1970) suggests that increased familiarity between the perpetrator of a crime and the complainant increases the likelihood that the situation will be handled informally. Black's analysis suggests that increasing inmate-inmate and inmate-staff familiarity under functional unit management will increase the informal handling of incidents and administrative remedies, despite the greater intensity of staff supervision.

The organization of the data in a weekly format presented some distinct disadvantages. Many variables that would contribute to the understanding of correctional climate simply do not occur frequently enough to be aggregated in this manner. Some of the variables considered, but eliminated for this reason, were suicide attempts, escapes, violations of furloughs, and attitude surveys (the CIES). Most of these variables were so rare that their occurrence could easily be influenced by factors completely outside the realm of institutional policy. The CIES survey administered in December 1976 exemplifies this problem. A day or two before the institution-wide administration of the survey, an inmate was shot trying to escape. The CIES was developed to measure overall institutional climate. However, institutional climate would be temporarily modified by an event such as this. Cavior (1977) notes that staff and inmate participation were perceptibly lower in the 1976 administration of the CIES than in earlier administrations, and the results of that survey indicated no measurable change in attitudes after the adoption of functional unit management. The post-FUM CIES results at Lompoc did not typify the pattern of improvement in climate reported at other institutions (Lee, 1980). A possible explanation for this irregularity was the untimely administration of the CIES near an unusual institutional event.

Method

The methodology consisted of intervention analysis in an interrupted time series design. The data will be analyzed for a three-year period, broken down in weekly increments, thus leading to approximately 156 observations for each variable under study. Approximately thirty weeks of data are available before the introduction of functional unit management (see Figure 1).

The particular method of time series modeling that was utilized in this study was the Box and Jenkins (1976) Autoregressive Integrated Moving

Figure 1. Time Line for Implementation of FUM at FCI, Lompoc

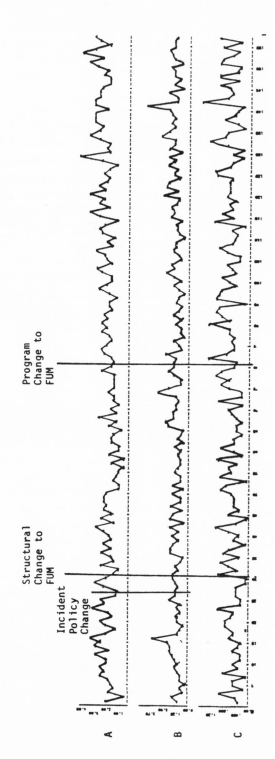

A = Less Serious Incident Rate
B = Serious Incident Rate
C = Administrative Remedy Rate

Average (ARIMA) model. The computer software used in the analysis was the package available through the Health Sciences Computing Facility of UCLA (see Liu, 1979). ARIMA models represent time series with three distinct components: (1) autoregressive, or those values systematically related to preceding values in the series (ϕ_n); (2) moving average, or those values systematically related to preceding error terms in the series (θ_r); and (3) dynamic trend or level components in the series (usually represented by the backshift operator, $1 - B^n$). Model identification is typically accomplished by visual and empirical analysis of the autocorrelation function (ACF) and the partial autocorrelation function (PACF). More thorough algebraic and substantive discussions of ARIMA modeling can be found in McCleary and Hay (1980) or, for the more mathematically sophisticated, in Box and Jenkins (1976).

Intervention analysis (Box and Tiao, 1975; Bhattacharyya and Layton, 1979) becomes relevant when we add an independent variable I_t to the model. In its simplest sense, the intervention factor can be thought of as equal to 0 prior to the point in time, t, in which the intervention took place and equal to 1 after that point in time. For instance, if an intervention takes place during the seventy-sixth week of the time series, then I_1 to $I_{75} = 0$ and I_{76} on $= 1$. This component (ω) is then added to the model and tested for significance using the Student's t-test.

Fortunately, the response to the impact or intervention can be interpreted more flexibly than a simple step function. The addition of a qualifying paramter (δ) to the 0–1 intervention component allows the investigator to test for the significance of interventions that are gradual at onset and permanent in duration, as well as those that are abrupt at onset and temporary in duration (McCain and McCleary, 1979, p. 62; McDowall and others, 1980).

The incident and administrative remedy data were obtained from the West Coast regional office of the federal prison system. Each case was individually identified by inmate number and date. The data was "cleaned" for out-of-range and missing values. The incident data were dichotomized according to seriousness level. All three data sets were then transformed from individual records to a count-by-week format. Inmate population figures for Lompoc were then extracted from data files in the Department of Justice computer. The counts were divided by the population figures and multiplied by one hundred to give an index of "rate per hundred inmates." This procedure has the effect of controlling for the inmate population at Lompoc, which varied by as many as two hundred over the time period under investigation. Starting January 1, 1975, the data reflect seven-day aggregate rates of each variable.

Each time series was subjected to the iterative ARIMA model-building strategy of identification, estimation, diagnosis, and metadiagnosis suggested by McCleary and Hay (1980). Once this process had eliminated the systematic relationships likely to be produced in time series data, the ARIMA component was identified and the interventions introduced into the model.

This analysis followed the suggestion of McDowall and others (1980, p. 84) and first tested an abrupt temporary impact for all independent variables. With the knowledge gained from that intervention test, the most accurate statistical form of the impact was determined.

For the incident data, the first intervention introduced reflected a July 7, 1975, alteration in incident handling policy. The August 1, 1975, change to the organizational structure of unit management and the July 7, 1975, change to the programming aspects of unit management were then tested. Since there was no direct change in policy relating to administrative remedies, that time series was tested with only the unit management interventions. The null hypothesis for these tests was that the intervention in question had no effect on the level of the time series being analyzed. The appropriate test of significance is the t-test.

Once the interventions were accepted or rejected, the total models were rediagnosed for their accuracy.

Results

This investigation studied the behavior of three separate time series (the rate of serious incidents per hundred inmates, the rate of less serious incidents per hundred inmates, and the rate of administrative remedies per hundred inmates). An initial visual inspection of all three raw series (Figure 1) suggests that they are all relatively stable both in level and in variance. They do not seem to reflect changes at the various posited intervention points.

An initial visual inspection of the ACF and PACF for the administrative remedy rate series suggests that differencing the series once would be appropriate (Figure 2). The first differenced ACF and PACF for the administrative remedy series suggests a first-order moving average component. When this component is entered into the model, θ_1 takes on the value of .9364 with a t-value of 36.76, clearly significant at the .05 level with 155 degrees of freedom. θ_1 is within the bounds of invertability (McCleary and Hay, 1980, p. 62). The resulting residuals from this model resemble white noise with a nonsignificant Q of 30.7 at the .05 level with 30 degrees of freedom. θ_0, or the mean of the once differenced series, is .0042 with a t-value of .0112, insignificant at the .05 level, and is dropped from the model. Thus, the ARIMA noise model for administrative remedy rates is

$$Y_t = \frac{1 - .9364B}{1 - B} a_t$$

Intervention testing for all series followed the general approach described below. All interventions were tested in the order in which they were introduced into the series, and if significant, retained in the model. For each

Figure 2. ACF and PACF for
Raw Administrative Remedy Rate Series

AUTOCORRELATIONS

```
            -1.0 -0.8 -0.6 -0.4 -0.2  0.0  0.2  0.4  0.6  0.8  1.0
            +----+----+----+----+----+----+----+----+----+----+
                                       I
  1   0.034                        +   IX  +
  2   0.008                        +   I   +
  3   0.129                        +   IXXX+
  4   0.118                        +   IXXX+
  5  -0.011                        +   I   +
  6   0.148                        +   IXXXX
  7   0.096                        +   IXX +
  8   0.089                        +   IXX +
  9   0.085                        +   IXX +
 10   0.107                        +   IXXX+
 11   0.120                        +   IXXX+
 12  -0.009                        +   I   +
 13  -0.008                        +   I   +
 14   0.123                        +   IXXX+
 15   0.043                        +   IX  +
 16   0.005                        +   I   + -
 17   0.036                        +   IX  +
 18   0.246                        +   IXXX+XX
 19   0.028                        +   IX  +
 20   0.080                        +   IXX +
 21   0.045                        +   IX  +
 22   0.041                        +   IX  +
 23  -0.085                        +  XXI  +
 24   0.103                        +   IXXX +
 25   0.083                        +   IXX +
 26   0.067                        +   IXX +
 27  -0.044                        +   I   +
 28   0.108                        +   IXXX +
 29  -0.021                        +   I   +
 30  -0.020                        +   I   +
```

PARTIAL AUTOCORRELATIONS

```
            -1.0 -0.8 -0.6 -0.4 -0.2  0.0  0.2  0.4  0.6  0.8  1.0
            +----+----+----+----+----+----+----+----+----+----+
                                       I
  1   0.034                        +   IX  +
  2   0.007                        +   I   +
  3   0.129                        +   IXXX+
  4   0.112                        +   IXXX+
  5  -0.019                        +   I   +
  6   0.135                        +   IXXX+
  7   0.064                        +   IXX +
  8   0.082                        +   IXX +
  9   0.059                        +   IX  +
 10   0.063                        +   IXX +
 11   0.098                        +   IXX +
 12  -0.058                        +   I   +
 13  -0.055                        +   I   +
 14   0.065                        +   IXX +
 15  -0.000                        +   I   +
 16  -0.013                        +   I   +
 17  -0.030                        +   I   +
 18   0.217                        +   IXXX+X
 19   0.018                        +   I   +
 20   0.061                        +   IXX +
 21  -0.019                        +   I   +
 22  -0.010                        +   I   +
 23  -0.099                        +  XXI  +
 24   0.024                        +   IX  +
 25   0.030                        +   IX  +
 26   0.046                        +   IX  +
 27  -0.085                        +  XXI  +
 28   0.041                        +   IX  +
 29  -0.086                        +  XXI  +
 30  -0.008                        +   I   +
```

intervention, the first test was a first-order pulse function. As McCleary and Hay (1980, p. 168) and McDowall and others (1980, p. 83) point out, the results of the first-order pulse function test, especially the value of the δ parameter, is likely to lead the investigator to the most accurate representation of the impact. Although there may be some a priori suggestion as to the form of the impact for all three of these series (for example, a first-order step function decreasing for administrative remedy rates when unit management is introduced), this suggestion is not strong enough to rule out rival tests of the form of the impact hypothesis.

For the administrative remedy rate series, the first intervention was at thirty weeks, when the organizational structure at Lompoc was officially changed to functional unit management. The introduction of the first-order pulse function yields the following parameter estimations:

$$\theta_1 = .9959 \ (t = 51.71)$$
$$\omega_0 = .2399 \ (t = 3.29)$$
$$\delta_1 = 1.0040 \ (t = 701.00)$$

Although the θ_0 parameter is significant and acceptable, the δ_1 value is outside the bounds of system stability (McCleary and Hay, 1980, p. 155). This high value for δ_1, however, implies that the effect is not damping out as would be expected in a pulse function. A zero-order step function is implied. The results of the zero-order step impact are:

$$\theta_1 = .9616 \ (t = 45.64)$$
$$\omega_0 = .1792 \ (t = 2.18)$$

The ω_0 value is significant beyond the .05 level with 154 degrees of freedom. This value can be translated to mean an increased level of approximately .18 administrative remedies per hundred inmates per week in the time period in which Lompoc changed its organizational structure to functional unit management.

The first-order step function was tested with the following results:

$$\theta_1 = .9646 \ (t = 47.33)$$
$$\delta_1 = .4475 \ (t = .49)$$
$$\omega_0 = .1094 \ (t = .60)$$

The impact parameters are insignificant and thus unacceptable. The most appropriate form of the intervention then, was the zero-order step function. This component will be added to the model.

During the eightieth week of the series, a new warden arrived and immediately began implementing many of the programs associated with functional unit management. The time period representing this intervention did not significantly affect the administrative remedy rate series. Although not statistically significant, an impact at period 80 worth noting was the zero-order step function. The w_0 parameter indicated an average increase in level of .122 administrative remedies per hundred inmates per week with a standard error of .079.

The final model for the administrative remedy rate series is representative of the increase in week 30 and the ARIMA noise model. The parameters of the modeling process are reflected in Table 1. The final algebraic model is:

$$Y_t = .1792(I_{30}) + \frac{1 - .9616B}{1 - B} (a_t)$$

The residuals of this model resemble white noise ($Q = 29.1$ with 28 degrees of freedom).

The ACF and PACF of the first differenced serious incident report series resembled those of the administrative remedy rate series and again suggested an ARIMA (0,1,1) model. The reader will note that differencing (subtracting the value of Y_{t-1} from Y_t for the entire series) has the effect not only of detrending the series, but of inducing a moving average component as well. As long as the moving average component is later modeled out to leave a cleaner error process, this procedure is advisable. The introduction of a first-order moving average component to the once differenced serious incident rate series yields the following parameters:

$$\theta_0 = .189 \quad (t = \quad .285)$$
$$\theta_1 = .8890 \quad (t = 24.340)$$

The trend component is insignificant and dropped from the model. The first-order moving average component is significant at the .05 level. The ACF and PACF of the error from this model indicate a white noise process with a Q of 21.0 at lag 30. The noise model for the serious incident rate series is represented by:

$$Y_t = \frac{1 - .8890B}{1 - B} (a_t)$$

For both incident rate series, the first proposed intervention was the change in incident policy at week number 26 in the series. If the intervention proved to be significant, it was kept in the model, thus controlling for any effect a change in policy had on incident or incident reporting behavior.

In general, none of the tests for impact in the serious incident series were significant at the .05 level, although there is some indication that there was a decrease in the rate of reported serious incidents at the program intervention (week 80). The null hypothesis that there was no change in the level of the serious incident rate at any of the three interventions is accepted.

Table 1. ARIMA Model and Impact Parameters for the Administrative Remedy Rate Series

Basic Statistics for the Series

Min = 0   Max = 1.41   Mean = .496   Standard Deviation = .284

Original ARIMA Noise Model.   ARIMA (0,1,1)

Source	Parameter	df
Model	θ_1 = .9364*	155
Model	θ_0 = .0042	156
Residuals	Q = 30.8	29

Tests for Intervention

Week Number	Type of Impact	ω_0 Value	t value	δ_1 value	t value
30	First order pulse	.2399	3.29*	1.0040[a]	701.00*
30	First order step	.1094	.60	.4475	.49
30	Zero order step	.1792	2.18*	-	-
80	First order pulse	-.1888	.68	.2010	.14
80	First order step	.0406	.48	.6891	.97
80	Zero order step	.1163	1.17	-	-

Final Noise and Impact model

Source	Parameter	df
Model	θ_1 = .9616*	154
Impact (week 30)	ω_0 = .1792*	154
Residuals	Q = 29.1	28

[a] beyond the bounds of system stability.

* significant at $\underline{p}$ < .05

After many different models were fit to the less serious incident rate series, the most efficient is an ARIMA $(0,1,1)_4$ model.

$$Y_t = \frac{1 - .6519 \, B^4}{1 - B^4} \, (a_t)$$

As with the serious incident reports, none of the forms of the impacts yields statistically significant results at weeks 26, 30, or 80.

Discussion

In light of the historical context of the development of functional unit management and the stated policy of the director of the federal prison system, it has been asserted that the primary goal of this management style is the establishment and maintenance of an organizational climate conducive to humane control. Although unit management was expected to contribute to voluntary participation in rehabilitative programs by inmates, the primary focus of this effort has been on the issue of humane control. Hence, the present study attempted to measure the impact of the implementation of unit management on the level of humane control in a federal correctional institution. The frequency of incident and administrative remedy reports was utilized as the operational indicator of the extent of humane control.

It was expected that a more personalistic approach to prison management would help fill the void in correctional goals left by disenchantment with the medical model of rehabilitation. Functional unit management's team-oriented, decentralized structure was expected to result in a significant reduction in the need for coercive psychological or corporal control of inmates. It was also expected to contribute to the management of and the stabilization of some previously informal patterns of authority and communication that exist in the prison setting.

An analysis of the data, using a simple interrupted time series quasi-experimental design, suggests that, in general, there were no changes in the dependent variables following the introduction of functional unit management at the Federal Correctional Institution at Lompoc, California. Clearly, these results are not an indication that functional unit management is not a worthwhile organizational style in the correctional setting, or even that it was not a success at FCI Lompoc. The flaws associated with drawing such unguarded and general conclusions from the results of this investigation are outlined below.

Goal Determination and Operationalization. As Weiss (1972) has pointed out, program goals are often hazy and ambiguous. A major part of the evaluator's task is to determine the exact nature of program goals, or to choose the most salient objective(s) from "a long list of pious and partly incomplete platitudes" (Weiss, 1972, p. 25). Glaser (1973) has also recognized the difficulty involved in translating officially mandated goals into unambiguous measures of performance. Functional unit management suffers from this official goal ambiguity. As reported earlier in the *Unit Management Manual* (1980), the federal prison system expected unit management to contribute to almost every phase of the correctional experience. By sifting through available evidence and putting unit management in historical perspective, this researcher has attempted to elucidate the objectives of functional unit management. However, it is apparent that in the process of focusing on humane control, some of the larger picture of FUM was lost.

Similarly, the attempt to operationalize humane control in terms of incident and administrative remedy reports is vulnerable to the criticism of overspecification. These two variables were used as behavioral measures of the concept of humane control. However, it is quite conceivable that humane control improved in the institutional setting while, for various reasons, the level of incident or administrative remedy reports remained unaffected. Given that this study was retrospective in its historical focus, attitudinal measures and other behavioral measures of humane control were unavailable. Future research in this area should seek to provide more insight into the measurable dimensions of humane control.

External Validity. Due to the constraints of data availability, only one institution was utilized in this research. Ideally, data would have been available from several institutions that had implemented unit management programs at different points in time. This approach would have made it possible to employ a more powerful design, such as Cook and Campbell's (1979, p. 213) "interrupted time series with switching replications." The advantages of this design are twofold. First, it provides for the introduction of a control group and thus approximates more closely a true experimental design. It also enhances external validity by sampling two subgroups of the population.

Even if this more rigorous approach were utilized, special caution regarding external validity in the institutional setting would be warranted. Prisons are unique places that tend to be unaffected by the generalizing influence of open society. Because of the uniqueness of individual institutions, the quantity or quality of behavior as well as the reporting of that behavior is highly dependent upon the particular institution. The confounding characteristic of the individuality of prisons makes interinstitutional comparisons difficult.

An additional threat to validity in correctional research is the prison system's jurisdictional barrier. The characteristics of federal, state, and local jurisdictions (such as the nature of the offender and the educational level or training of staff) may vary to a large degree. These characteristics should be taken into consideration when generalizing the results of an investigation to all prison settings.

Organizational Change. In lieu of specific evidence to the contrary, this study has assumed that functional unit management was fully implemented on the dates represented by official actions. The administrative, authoritative, and program structures did indeed begin on the announced dates. However, one of the primary intermediate objectives, inmate-staff familiarity, which was expected to contribute to achieving the ultimate goal of enhanced humane control, is difficult to isolate in terms of time. To determine exactly how familiar staff and inmates must be with each other in order to create a more open and trusting atmosphere is problematic. Whether or not familiarity, as an intervening process, enhances humane control is itself a valid research question. Future evaluations will do well to test assumptions concerning mediating processes before drawing any definitive conclusions.

It may also be that data aggregation by weeks represents too fine a distinction for testing organizational change. Monthly data may be more appropriate for testing variations in organizational structure. Again, limitations on the availability of data made monthly aggregation infeasible for this particular investigation.

Conclusion

This investigation makes the following significant contribution to the evaluation of functional unit management as an organizational style. First, it presents an organizational, historical, and theoretical framework for quantitatively testing the success of unit management. This framework, although it may not reflect all aspects of a multifaceted organizational structure, provides future evaluators with an analytical context within which to work. Second, the quasi-experimental design known as an interrupted time series is exemplified. This research paradigm helps to eliminate many of the methodological problems associated with measuring change over time.

Finally, the results derived from this research are informative at their face value. Despite some of the problems associated with drawing general conclusions from these findings, the results indicated that the change to unit management was accompanied by no significant change in incident reporting rates and only a slight increase in administrative remedy rates at the structural change to unit management. The increase in administrative remedy rates is counter to the direction of change that unit management was expected to produce. In retrospect, it is possible that increased staff availability and more inmate-related decisions made by staff caused the number of complaints about those staff and their decisions by inmates to increase. However, the general finding of "no effect" by unit management on two important indicators such as incident reports and administrative remedy reports is an indicator that unit management, or its implementation at Lompoc, may not be living up to its full potential.

References

Bhattacharyya, M. N., and Layton, A. P. "Effectiveness of Seat Belt Legislation on the Queensland Road Toll—An Australian Case Study in Intervention Analysis." *Journal of the American Statistical Association,* 1979, *74* (367), 596–603.

Black, D. J. "Production of Crime Rates." *American Sociological Review,* 1970, *35* (4), 733–748.

Bogan, J., Karacki, L., and Lansing, D. "Evaluation of Functional Unit Management." Unpublished report to the Federal Prison System, Washington, D.C., September 1975. Mimeographed.

Box, G. E. P., and Jenkins, G. M. *Time Series Analysis: Forecasting and Control.* (Rev. ed.) San Francisco: Holden-Day, 1976.

Box, G. E. P., and Tiao, G. C. "Intervention Analysis with Applications to Economic and Environmental Problems." *Journal of the American Statistical Association,* 1975, *70,* 70–79.

Carlson, N. A. "Corrections in the United States Today: A Balance Has Been Struck." *The American Criminal Law Review,* 1976, *13* (4), 615–647.

Cavior, H. "Results of the Correctional Institutions Environments Scale at FCI and FPC, Lompoc," Unpublished report to the Federal Prison System, Washington, D.C., July 1977. Mimeographed.

Cook, T. D., and Campbell, D. T. *Quasi-Experimentation: Design and Analysis for Field Settings.* Chicago: Rand McNally, 1979.

Fogel, D. . . . *We Are Living Proof. . . The Justice Model for Corrections.* Cincinnati: W. H. Anderson, 1975.

Glaser, D. *Routinizing Evaluation: Getting Feedback on Effectiveness of Crime and Delinquency Programs.* Rockville, Md.: National Institute of Mental Health, 1973.

Lansing, D., Bogan, J., and Karacki, L. "Unit Management: Implementing a Different Correctional Approach." *Federal Probation,* 1977, *41* (1), 43–49.

Lee, P. "Overview: The Social Climate Effects of Functional Unit Management." Unpublished report to the Federal Prison System, Washington, D.C., August 1980. Mimeographed.

Lejins, P. P. *Criminal Justice in the United States 1970–1975.* College Park, Md.: The American Correctional Association, 1975.

Liu, L. *Users Manual for BMDQ2T: Time Series Analysis.* Technical Report No. 57. Department of Bio-Mathematics, University of California, 1979.

McCain, L. J., and McCleary, R. "The Statistical Analysis of the Simple Interrupted Time-Series Quasi-Experiment." In T. D. Cook and D. T. Campbell (Eds.), *Quasi-Experimentation: Design and Analysis for Field Settings.* Chicago: Rand McNally, 1979.

McCleary, R., and Hay, R. A., Jr. *Applied Time Series Analysis for the Social Sciences.* Beverly Hills: Sage, 1980.

McDowall, D., McCleary, R., Meidinger, E. E., and Hay, R. *Interrupted Time Series Analysis.* Beverly Hills: Sage, 1980.

Moos, R. H. *Evaluation Correctional and Community Settings.* New York: Wiley, 1975.

Rowe, R., Foster, E., Byerly, K., Laird, N., and Prather, J. "The Impact of Functional Unit Management on Indices of Inmate Incidents." Unpublished report to the Federal Prison System, Washington, D.C., August 1976. Mimeographed.

Unit Management Manual. Washington, D.C.: Federal Prison System, 1980.

Webb, E. J., Campbell, D. T., Schwartz, R. D., and Sechrest, L. *Unobtrusive Measures: Nonreactive Research in the Social Sciences.* Chicago: Rand McNally, 1966.

Weiss, C. *Evaluation Research: Methods of Assessing Program Effectiveness.* Englewood Cliffs, N.J.: Prentice Hall, 1972.

Michael Janus is a research analyst with the federal prison system in Washington.

Data routinely collected for monitoring purposes may often be useful for exploring causal impacts of program changes.

Using Routine Monitoring Data to Identify Effects and Their Causes

Roger B. Straw
Nancy M. Fitzgerald
Thomas D. Cook
Stephen V. Thomas

Interrupted time series analysis (ITS) has traditionally been used in two related situations. The first, and perhaps the most common, use of ITS is to evaluate how large scale social or physical changes affect aggregate indicators, such as how television influences crime rates (Hennigan and others, 1982) or how drought and conservation campaigns influence water consumption (Agras and others, 1980). In such cases, the data are usually aggregated over months or years and over such large populations as cities or nations. The second use of ITS is to investigate the effects of planned interventions in relatively small populations over relatively short periods of time. For example, studies have examined the effects of educational interventions on school-level dropout rates (Gottman and McFall, 1972) and on students' disruptive behavior (Hall and others, 1971). Some single-subject research in behavior therapy is also of this second type. Both types of ITS depend on relatively long time series (fifty or

G. Forehand (Ed.). *New Directions for Program Evaluation: Applications of Time Series Analysis to Evaluation*, no. 16. San Francisco: Jossey-Bass, December 1982.

more observations) and an intervention with an abrupt and known onset that should cause either an immediate change in the dependent series or a delayed change whose temporal lag has been specified in advance.

In this chapter, we seek to generalize the logic of ITS methodology to make it relevant to situations (1) where the emphasis is on identifying effects and then discovering their cause, rather than on probing for the effects of a given potential cause; (2) where many individual time series exist of only moderate length (twenty or more observations); and (3) where the onset of the intervention is not known. Our goal is first to identify individual cases where an effect can be inferred from a shift in the time series, and then to explore what may have caused the shifts that have been identified. The data sets to which such an approach is applicable are routinely produced by management information systems in government or private industry.

This approach is conceptually similar to cross-sectional methods used in education to identify superior schools based on some criterion measure, such as student-body achievement scores (Clark, Lotto, and McCarthy, 1980; Edmonds, 1979). After such schools have been identified, an examination of program-related variables is then carried out in an attempt to identify manipulable and transferrable causal factors that other schools might emulate in order to increase their own effectiveness. A major problem with this cross-sectional approach is that schools may be spuriously identified as superior on the basis of "outlier" performance based on scores that have been disproportionately influenced by chance. An advantage of time series methodology is that single deviant scores need not be used for categorizing entities as superior or successful. Instead, the categorization can be based on discontinuities in a time series that persist for a specified period of time, and so are less likely to be due to chance.

In many ways, the approach we are proposing represents a reaction to the many no-difference findings in past evaluations of local projects. Such no-difference findings are often uninterpretable, because we cannot be sure of the reasons for the observed lack of statistically significant effects. In some cases, the theory behind the project may be inadequate or wrong. But in other cases, the theory may be fine but may have been poorly captured by the project as it was actually implemented. Another possibility is that the evaluation itself may have been conducted inadequately and may have failed to detect true effects. The present approach, based on detecting cases where change has demonstrably occurred, shifts the basic evaluation question from: "What is the effect of X (a program or project) on Y (an outcome or set of outcomes)?" to "Given an already observed effect on Y, what changes in X seem to have caused it?"

This change in the form of the primary evaluative question also suggests a model of evaluation that differs substantially from traditional models, most of which are concerned with judging the effectiveness of a program either in the context of its future retention (Campbell, 1969), as a test of a theoretical program model (Rossi, Freeman, and Wright, 1979), or as a test of procedures for making improvements within a program or local project (Cronbach and others, 1980). A model based on detecting outliers, using some criterion of successful outcome, suggests an approach in which a program is conceived as a number of individual projects that are implemented in different ways at different sites at different times, and that this variability in implementation may be related to the degree of effectiveness. The goal of evaluation under this model is to identify those local projects funded out of a general program that are effective, and to explore these projects to detect possible similarities in implemented activities that might account for their effectiveness. The exploration for potential causes of observed effects forms the heart of the approach.

The remainder of the chapter will describe the generic methodology and present an illustration of its application. The methodology has three stages: (1) selecting sites with deviations in their time series data that one is willing to infer are "effects"; (2) collecting information on factors that might be causally related to the detected changes; and (3) relating changes in the potential causal variables to the changes observed in the time series. When a reasonably large number of cases is available, a fourth stage can be added, which involves conducting confirmatory analyses of the potential causal relationships identified in the third stage. The example presented in this chapter is taken from an evaluation of student participation in the National School Lunch Program in the state of New York. The time series data are participation rates over a thirty-three-month period for over eight hundred individual elementary schools.

Stage 1: Identifying Deviations in Individual Time Series

Method.

The identification of deviations begins by describing the typical pattern of variability across the total sample of time series. This creates a standard or norm against which to identify individual schools with unique changes during the data collection period. To develop a normative composite, we averaged the data across all the schools for each time point. Since the data were monthly, we therefore calculated the average value for month 1, month 2, month 3, and so forth. This allowed us to describe the average series in terms of its level, trend, and seasonality.

Such a composite can be useful in its own right, since examination of seasonal patterns or time trends can provide program staff with important insights about operations or changes that may be needed in routine management. For example, a plot of the composite time series for the New York schools indicated that participation by children who obtained lunches free or at reduced cost was unusually low in September of each school year, perhaps because of delays in determining if students were eligible for free or reduced-price meals. Modifying the starting date of eligibility screening or the length of time screening takes might permit more children from poorer homes to eat a free or reduced-price lunch earlier in the school year.

For the present purpose, the composite time series is more important in providing a standard against which to compare each of the more than eight hundred time series from individual schools. The goal is to identify time series with discontinuities that are different from any discontinuities found in the composite, for these will be treated as effects that, upon further exploration, may reveal unique changes in program activities at particular sites that can be transferred elsewhere. Although individual series may differ from the composite in many ways, some differences are of little interest. For instance, stable differences in the level of participation in the school lunch program are likely to reflect nonmanipulable, structural differences between schools, such as differences in the average poverty level of children. The discontinuities of most concern to us are abrupt changes in level or slope that occur within a time series and persist for a reasonable length of time. These are more likely to result from changes in variables that are potentially manipulable and transferrable, and are therefore of particular relevance for policy.

The most basic approach to identifying discontinuities involves the visual inspection of each time series. Plots of individual series can be generated with the basic data untransformed. However, these plots will often be difficult to interpret because of the complex patterns created by seasonal and other variations and by random variation that will be especially pronounced in individual cases when compared to the composite.

Thus, it is useful to remove seasonal patterns and overall trends before generating the individual scatterplots for visual inspection. This can be accomplished by computing a new time series for each site in which each observation represents the difference between the raw data and the corresponding mean value across all of the individual series (that is, the composite). For any site that is exactly like the overall composite, this procedure will result in a horizontal line at the zero point of the Y-axis. For any site and using deseasonalized data, we define a discontinuity as an abrupt change in level or slope that occurs sometime after the first observation.

Visual inspection is useful for making the analyst aware of the nature of the data, for corroborating statistical analyses, and for persuading readers whose intuition depends on striking graphic evidence rather than complex, and perhaps unfamiliar, statistical procedures. However, visual inspection has been criticized as unsystematic and as impractical when many time series are involved. Our experience with the school lunch data suggests that neither of these arguments is absolutely true. While visual inspection of a large number of time series is certainly tedious, it is nonetheless feasible, provided that the analyst is interested only in detecting changes that are large enough to be visually striking. Our experience indicates that visual inspection can be accomplished relatively quickly and with acceptable levels of reliability.

On the other hand, statistical methods for studying individual time series will probably become increasingly important as more agencies regularly collect time series data for monitoring purposes. If our approach is to be widely applicable, it will require the development of an efficient statistical method for detecting outlier time series, whether these be unique average values across the whole series (for example, levels or slopes) or unique changes within a series (for example, abrupt discontinuities in level or slope).

In the process of conducting the evaluation of the school lunch participation data, we examined several possible methods and have identified some promising alternatives (Cook and others, 1980). The most promising was based on an ordinary least squares regression analysis of the deviation scores (raw score minus composite score) standardized across schools. The average level and slope of the series for one six-month interval was compared with the average level and slope over the next six months. So, for example, months 1 through 6 were compared to months 7 through 12, months 2 through 7 to months 8 through 13 and so on. Differences in average levels and slopes between six-month intervals were tested for significance using adjustments for multiple comparisons. Work is continuing on more refined methods that better control for Type 1 error problems associated with correlated error and other characteristic features of time series data (Thomas, 1982).

The test of the utility of this (or any) statistical procedure is two-fold. First, it should be more time efficient than the alternative. In this case, many time series can be described in considerably less time than it takes to generate and visually inspect scatterplots. Second, the statistical procedure should allow the investigator to identify effects that are difficult or impossible to detect with the eye. Whether it does this is an empirical issue, and so a comparison of the statistical procedure with visual inspection is needed. One such comparison is presented in the next section.

Example: Time Series Data on School Lunch Participation

Definition of Participation. Monthly data were obtained from New York State records for the period September 1976 to November 1979. Given school years of ten months, this meant thirty-three time points. A total of 859 public and private schools comprising 248 food-service districts were involved. All were elementary schools serving kindergarten or first grade through sixth grade. Schools from New York City were excluded from the study because they do not report school-level data to the state. For each school in the sample, the following were reported on a month-by-month basis: student enrollment, average daily attendance, the number of students eligible for free and for reduced-price lunches, the number of days lunch was served, and the number of lunches served by payment category (that is, free, reduced, and full).

From these data, monthly participation rates were calculated for the three payment categories and for total participation. The numerator in each case was the total number of lunches served in a particular category. The denominator was the number of lunches that would have been served had all students eligible in a particular category eaten lunch on every day that it was served. The denominators were estimated by multiplying the number of days lunch was served in the month by the number of students eligible for that type of lunch. So, for example, total participation in September was obtained by the formula:

$$\frac{\text{Total lunches served in September}}{\text{Number of days lunch served in September} \times \text{Total enrollment in September}}$$

The measure reflects the proportion of times that the average student eats lunch during the month. It will change if either the number of participants changes or if the participants change how often they eat. However, it does not allow us to specify to what extent changes are due to one or the other of these factors.

State-Wide Composites. After calculating the participation rates for each of the payment categories and also for total participation, we simply plotted state-wide averages of participation rates over the thirty-three months of the study in order to describe the state-wide patterns of participation in terms of level, trend, and seasonal pattern.

The results of this procedure are presented in Figure 1. The overall level of participation was highest in the free and reduced categories (about 84 percent and 75 percent, respectively) and lowest in the full category (about 40 percent). Participation in the free and reduced programs remained constant, but participation in the full payment program increased during the time of this study, from an average of about 37 percent in the first year to about 42 percent in the third year. The seasonal pattern was basically the same each year for all three payment categories: participation increased from September to December, decreased from December to February, increased slightly in March and April, and then dropped off in May and June. The increase from September to October was particularly striking for the free and reduced categories. This may have been because eligibility was not determined until after the school year began in September. Determining eligibility sooner *might* increase the September participation rates.

Identifying Time Series with Discontinuities. In this phase, we plotted the

participation data by payment category for each individual school as deviations from their respective state-wide composites presented in Figure 1. Discontinuities of two basic types were examined: (1) abrupt changes in the mean
level of the participation series (for example, when participation suddenly
increased or decreased, such as ____⌐‾ or ‾⌐____ and (2) abrupt

**Figure 1. Statewide Composites of
Participation Rates by Payment Category**

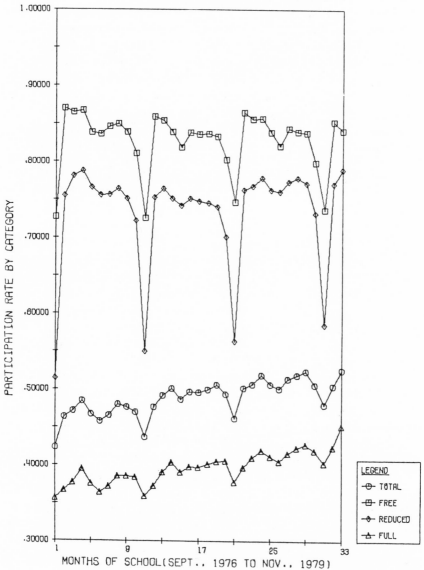

PARTICIPATION RATE BY CATEGORY

MONTHS OF SCHOOL(SEPT.. 1976 TO NOV.. 1979)

LEGEND
-⊖- TOTAL
-⊞- FREE
-◆- REDUCED
-△- FULL

changes in slope where participation began to increase or decrease at a faster rate than it had before (for example, ——————✓ or ———————✗). Only discontinuities of six months or longer duration were considered as meaningful effects and retained for further study. As mentioned previously, it was easier than one might expect to identify these types of discontinuities using visual inspection. Of the 175 randomly chosen schools examined for exploratory purposes, we agreed on the presence or absence of a discontinuity in 88 percent of the individual time series. (This was agreement on 626 of the 700 free, reduced, full, and total participation plots that were generated for the subset of 175 schools.) A total of 90 schools (51 percent) showed a discontinuity of sufficient magnitude and duration to be examined further.

The regression-based statistical technique that we developed, although preliminary, proved to have a number of strengths. It detects smaller effects, which may be advantageous in exploratory research. It is also less susceptible to fatigue, distraction, or sources of arbitrariness other than those programmed in. Moreover, the accuracy of the statistical procedure is particularly noteworthy in cases where more than one effect occurs within the same series.

Stage 2: Collect Information on Potential Causal Variables

Method

Once the sites have been identified at which a change in the dependent variable occurred, the next step is to generate a list of potential causes and to discover if any changes occurred in them shortly before the observed effect occurred. The list of potential causes is generated by using relevant theory and experience. The emphasis should be on variables presumed to be transferrable or manipulable and variables that are presumed to be necessary for the emergence of the effect in question. The list of potential causes will necessarily be imperfect, incomplete, and may be long.

Once the list has been constructed, information is needed on which variables changed during the period of the time series. Data can usually be obtained from two sources. The first is archival and can involve records similar to those from which the original time series data were obtained. Such data will usually cover the same time period, and may even be available on the same cases and with the same time intervals as the data on effects. If so, the evaluator will not have to depend on the memory of individuals about what changes occurred at particular times in particular sites. Unfortunately, the information so archived will usually be limited to a small subset of all the potential causes on the list.

An alternative approach is to interview persons at the particular sites with discontinuities in order to collect retrospective data on what changed at the time the effect was observed. If this method is used, care must be taken to ask the questions so that the information is in usable form. For example, it is crucial that the timing of a reported change in some potential cause be specified as exactly as possible. Knowing that the price of a commodity used in the school lunch program changed sometime in 1977 may not be specific enough if the time series data is based on monthly consumption data. The evaluator should also use this opportunity to have respondents suggest potential causes in addition to those already generated. Finally, if information on some variables can be obtained from multiple sources, the overall quality of any one source of information can be assessed and appropriate adjustments made, because of the triangulation.

Once the list of potential causes has been developed, exploratory analyses can then be conducted to identify a subset of variables for further analysis. First, a number of variables can usually be eliminated from the list of potential causes because they changed at so few of the sites. While the approach we are describing can in theory be used in a case study of only one site, most of the analyses we propose depend on a number of sites having instituted a particular change. This permits data to be aggregated across these sites, thereby producing a more stable aggregate series. Second, for each possible cause a simple cross-tabulation of sites should be conducted to relate those with changes to those having discontinuities in their time series. Variables on which sites report changes but no discontinuities can be eliminated from further analysis. What remains, hopefully, is a manageable subset of explanatory variables on which project sites with time series discontinuities are known to have changed. These are the prime candidates for further analyses.

Example

A literature review and conversations with state-level officials led to a list of ninety-two constructs that could potentially influence participation in the school lunch program. These were incorporated into a questionnaire that school food-service directors completed for the schools in their districts. (For districts with more than three schools, the director was asked about a random sample of three of them.) The questionnaire solicited responses to questions about changes in the ninety-two possible causes during the previous three school years.

Seventy-three percent of the questionnaires were returned, with only minimal telephone encouragement to do so. We began our analysis by estimating which variables could *not* have been systematic causes of the observed discontinuities because (1) schools rarely changed that particular aspect of their lunch program (that is, the potential cause changed in fewer than five of the 175 schools initially being studied) and (2) a reasonable number of schools showed changes on a potential cause but they were not the same schools where discontinuities in participation were observed. These critiera reduced the number of potential causes from the initial ninety-two to thirty-three. This latter group included: changes in the price of lunch in the full and reduced categories, changes in student involvement in menu-planning activities, changes in lunchroom atmosphere, and many other factors.

However, thirty-three potential causes were still considered too many. So we used two additional techniques to eliminate potential causes. First, changes on the variables were classified as increases or decreases and then related to the type and direction of the observed discontinuity in the observed time series. Nine additional variables were dropped at this point because insufficient numbers of sites consistently increased or decreased, or because the direction of the changes was not related in any systematic way to the direction of changes in the time series. Second, an exploratory factor analysis was undertaken to determine if some of the variables could be combined into broader constructs. As a result of the analysis, four constructs were formed: menu planning, food options, nutrition education, and food quality. Twelve additional variables were also retained including price, lunchroom atmosphere, and others.

Stage 3: Relating Changes in the Program to Changes in the Time Series

Method

Once a subset of the potential causal variables has been identified that changes on a reasonably large numer of sites and seems related to observed effects in the time series data, a more refined analysis can be conducted of the relationship between observed effects and potential causes. Where the timing of the change in the independent variable is well specified and is expected to result in an immediate, abrupt change in the dependent variable time series, an ITS approach may be used. If the individual time series are long and stable, the original data for each case may be analyzed separately using traditional autoregressive integrated moving average (ARIMA) modeling. In the likely event that the individual series are short and unstable, composites formed by averaging the time series of sites that report similar changes in the causal variable (for example, decreases in price) at the same time can be examined instead. Control time series can be formed similarly from sites that report no changes on the causal constructs being investigated.

When the available information is not specific enough to allow the evaluator to use an ITS approach, a second wave of data collection may be required. This group of interviews can be limited to just those sites and potential causes already identified as having concurrently changed. At this stage, the questions can be much more specific and tailored to the individual site. However, if additional data collection is not possible because of limited resources or time, alternative design strategies may be necessary. For example, when a number of sites change on the same potential cause at roughly the same time, they can be grouped together and compared either to a noncomparable control group of sites reporting no change on the potential cause or to sites reporting the same change but at a different point in time. But because the sites in any one group change at somewhat different times, the data may have to be collapsed across time intervals, perhaps resulting in nothing finer than a single pretest and a single posttest mean. Obviously, the internal validity and statistical power of such an analysis can be seriously compromised if the information on the time of intervention is inaccurate.

The actual analysis of the quasi-experimental data resulting from the above approaches may be either statistical or more qualitative. In the case of ITS designs, both the treatment and the control series can be modeled using ARIMA methods and tested for the presence of an intervention component. Alternatively, the two series can be analyzed visually to see if there is any clear evidence of a greater change after the potential cause than before it. If ITS analysis is not possible and fallback options involving simple pretest-posttest designs have to be resorted to, the analysis can once again be either statistical or visual. Both the ITS and pretest-posttest approaches are illustrated in the following section that relies heavily on statistical analysis. However, we suspect that many situations to which this generic approach can be applied will require a qualitative type of analysis rather than a statistical one.

Example

Because of limitations in the questionnaire design, we had data on the year when schools changed on any causal variable, but we did not have the specific date within the year when the change occurred. Thus, we were able only to establish whether a potential cause changed from one year to the next, the consequence being that all changes had to be treated as if they had occurred between school years. Because of this, a quasi-experimental strategy was developed that used average yearly participation rates in the context of a pretest-posttest control group design. The purpose was to examine each of the potential causes identified in Stage 2. Changes in the price of full payment lunches will be presented as an example because it is a variable for which the assumption of between-year change could be reasonably justified (Hansen, 1980).

Four quasi-experimental designs were used to examine the effects of increases in the price of full price lunches. In all four cases, the treatment group was composed of schools that increased their prices. Design 1 used a control group composed of schools that reported no changes on any of the major potential causes. Design 2 compared a group of schools that increased price in one year with a group of schools that changed their price the next year.

Designs 3 and 4 compared schools that changed their prices in 1977–1978 or 1978–1979, respectively, with schools that did not change their price at all in any year but that did change on other potential causes. These particular analyses were not restricted to the 175 schools whose plots we visually inspected but included any of the schools in our total sample with reliable participation data.

For illustrative purposes, the results of Design 3 are presented in Table 1. The 47 schools in the "increase" group reported increases in full price between the 1977–1978 and the 1978–1979 school years. The 246 schools in the "no change baseline" group did not change their full price between those two school years, though they did change on at least one other potential cause. The average participation rates for each school year are reported. In the control group there is a steady increase in participation across the years but this increase is not apparent after the price increase in the group of schools that raised prices (ANCOVA; $F(1,290 = 6.26, p < .01)$. With all four of the designs, the results were similar, suggesting that the effect can be found across three different control groups (Designs 1 through 3) and at two different times of change (Designs 3 and 4). In addition, participation among students in the free and reduced-price lunch categories was not affected at schools with increases in the full price. This makes it unlikely that some school or program-related variable caused the decline we here attributed to price increases.

The same quasi-experimental approach was followed for the other identified potential causes, but only the data on price changes were analyzed statistically. None of the other potential causes had strong or consistent effects on mean participation rates. However, we want to warn the reader that with all variables other than price, we are unsure when the change took place. If it took place during a school year, then the current method would be less likely to detect true effects, for some pretreatment months would be in the posttest year or some posttreatment months in the pretest year. This illustrates just how crucial it is to know precisely when changes occurred in potential causes.

Summary

A method was described for using routinely collected data from large management information systems for program development, and an example

Table 1. The Effects of Price Increases on Participation Rates in the Full-Payment Category

	School Year			
Schools with	*1977–1978*	*1978–1979*	*Intervention*	*1979–1980*
Price Increases (*N* = 47)	41.0%	41.9%	Price Increase	41.8%
No Change (*N* = 246)	40.0%	41.6%	No Change	43.5%

of the method was presented from the school lunch program in New York State. The approach is intended primarily to allow an evaluator to explore potential causes for observed changes in performance at the various project sites where a program is being carried out. As such, it represents a substantial departure from previous models of evaluation that begin by assuming they know the potential cause of interest (that is, a program) and then set out to discover its effects. In the present approach, the evaluator first sets out to discover effects and only then goes in search of their causes. As we have outlined it here, the approach is most sensitive for identifying the subset of causes that have an immediate, abrupt impact on the dependent variable. However, with sufficient background knowledge about the temporal lag between cause and effect, any potential cause could be investigated.

The example of school lunch participation illustrates many of the strengths of the approach, particularly with respect to the use of school-level time series data to detect sudden changes in the level or slope of participation that persisted for at least six months. However, the example also illustrates the imperfect and often inconclusive nature of attempts to explore the causes of such effects. In the present case, the exploration of potential causes and their relationship to observed effects could have been greatly enhanced by obtaining information about the exact month when variables changed so as to improve either ITS or other quasi-experimental analyses of the causal efficacy of presumed causes. Also, telephone or on-site interviews would have been useful, particularly in those sites where a change had been detected and could be attributed to a certain month. The interviews could then have involved probes of all the events that occurred at that time, with careful attention being paid to eliciting more than one source of data that converged on a particular change at a particular date. Attempts at convergence are all the more important since the search for causes is retrospective and distortions of memory are possible. Our guess at this time is that the search for causes will often be qualitative, even if the identification of effects is not. In the present example, both stages were quantitative, and had we wanted we could have analyzed both stages as abbreviated interrupted time series instead of analyzing only the first this way. It is not yet clear how typical our quantitative example will turn out to be; qualitative analyses may be more common than quantitative ones.

Sometimes, it will be possible to divide a sample of sites into two sets, one of which is analyzed in Stage 3 to explore causation, and the other of which is used later to confirm any results of Stage 3. This desirable state of affairs only makes sense when data are available from many sites, and how often this occurs is not at all clear. Our provisional guess is that many management information systems contain data that are routinely collected from a large number of sites for monitoring purposes, including government agencies in education, mental health, occupational safety, civil rights, and so on. If so, it may not be too difficult to identify sites where beneficial changes have suddenly occurred and to probe why they came about. If this is indeed possible, it should add an important tool to the current armamentarium for improving national programs and to the local projects that comprise these programs.

References

Agras, W. S., Jacob, R. G., and Lebedeck, M. "The California Drought: A Quasi-Experimental Analysis of Social Policy." *Journal of Applied Behavior Analysis*, 1980, *13*, 561–570.

Campbell, D. T. "Reforms as Experiments." *American Psychologist*, 1969, *24*, 409–429.

Clark, D. L., Lotto, L. S., and McCarthy, M. M. "Factors Associated with Success in Urban Elementary Schools." *Phi Delta Kappan*, 1980, *61*, 467–470.

Cook, T. D., Straw, R. B., Fitzgerald, N. M., Thomas, S., Cook, F. L., Ferb, T. E., Magidson, J., Napior, D., and St. Pierre, R. G. *Assessing Determinants of Participation in the School Lunch Program in New York State from 1976 to 1979*. Cambridge, Mass.: Abt Associates, 1980.

Cronbach, L. J., Ambron, S. R., Dornbusch, S. M., Hess, R. D., Hornik, R. C., Philips, D. C., Walker, D. F., Weiner, S. S. *Toward Reform of Program Evaluation: Aims, Methods, and Institutional Arrangements*. San Francisco: Jossey-Bass, 1980.

Edmonds, R. "A Discussion of the Literature and Issues Related to Effective Schooling." Unpublished manuscript, ERIC Clearinghouse, 1979.

Gottman, J. M., and McFall, R. M. "Self-Monitoring Effects in a Program for Potential High School Dropouts: A Time Series Analysis." *Journal of Consulting and Clinical Psychology*, 1972, *39*, 273–281.

Hall, R. V., Fox, R., Willard, D., Goldsmith, L., Emerson, M., Owen, M., Davis, S., and Porcia, E. "The Teacher as Observer and Experimenter in the Modification of Disputing and Talking-out Behaviors." *Journal of Applied Behavior Analysis*, 1971, *4*, 141–149.

Hansen, S. J. *Impact of Changes in School Lunch Prices on Pupil Participation*. Washington, D.C.: American Association of School Administrators, 1980.

Hennigan, K. M., Del Rosario, M. L., Heath, L., Cook, T. D., Wharton, J. D., and Calder, B. J. "The Impact of the Introduction of Television on Crime in the United States: Empirical Findings and Theoretical Implications." *Journal of Personality and Social Psychology*, 1982, *42* (3), 461–477.

Rossi, P. H., Freeman, H. E., and Wright, S. R. *Evaluation: A Systematic Approach*. Beverly Hills: Sage, 1979.

Thomas, S. V. "Statistical Methods for Identifying Discontinuities in Time Series Data." Unpublished doctoral dissertation, Northwestern University, 1982.

Roger B. Straw is a health planner with Piedmont Heatlh Systems Agency in Greensboro, North Carolina, and a clinical instructor of psychology at the Bowman Gray School of Medicine.

Nancy M. Fitzgerald is a postdoctoral fellow in social psychology at Northwestern University in Evanston, Illinois.

Thomas D. Cook is a professor of psychology at Northwestern University, Evanston, Illinois.

Stephen V. Thomas is completing his doctorate in statistics from Northwestern University, Evanston, Illinois, and is employed as a statistical methods analyst with UOP Process Division in Des Plaines, Illinois.

Index

Statement of Ownership , Management, and Circulation
(Required by 39 U.S.C. 3685)

1. Title of Publication: New Directions for Program Evaluation. A. Publication number: 449-050. 2. Date of filing: 9/30/82. 3. Frequency of issue: quarterly. A. Number of issues published annually: four. B. Annual subscription price: $35 institutions; $21 individuals. 4. Location of known office of publication: 433 California Street, San Francisco (San Francisco County), California 94104. 5. Location of the headquarters or general business offices of the publishers: 433 California Street, San Francisco (San Francisco County), California 94104. 6. Names and addresses of publisher, editor, and managing editor: publisher—Jossey-Bass Inc., Publishers, 433 California Street, San Francisco, California 94104; editor—Scarvia B. Anderson, ETS, 250 Piedmont Avenue NE, Atlanta, Georgia 30308; managing editor—William E. Henry, 433 California Street, San Francisco, California 94104. 7. Owner: Jossey-Bass Inc., Publishers, 433 California Street, San Francisco, California 94104. 8. Known bondholders, mortgages, and other security holders owning or holding 1 percent or more of total amount of bonds, mortgages, or other securities: same as No. 7. 10. Extent and nature of circulation: (Note: first number indicates average number of copies of each issue during the preceding 12 months; the second number indicates the actual number of copies published nearest to filing date.) A. Total number of copies printed (net press run): 3745, 3687. B. Paid circulation, 1) Sales through dealers and carriers, street vendors, and counter sales: 85, 40. 2) Mail subscriptions: 1701, 1529. C. Total paid circulation: 1786, 1569. D. Free distribution by mail, carrier, or other means (samples, complimentary, and other free copies): 125, 125. E. Total distribution (sum of C and D): 1911, 1694. F. Copies not distributed, 1) Office use, left over, unaccounted, spoiled after printing: 1834, 1993. 2) Returns from news agents: 0, 0. G. Total (sum of E, F1, and 2—should equal net press run shown in A): 3745, 3687. I certify that the statements made by me above are correct and complete.

JOHN R. WARD
Vice-President

DATE DUE